Renato Lemos Santos
Monalisa B. Costa-Santos

Biological nitrogen fixation in sugarcane

Renato Lemos Santos
Monalisa B. Costa-Santos

Biological nitrogen fixation in sugarcane

History, concepts, evaluation methods and sources of variation

ScienciaScripts

Imprint
Any brand names and product names mentioned in this book are subject to trademark, brand or patent protection and are trademarks or registered trademarks of their respective holders. The use of brand names, product names, common names, trade names, product descriptions etc. even without a particular marking in this work is in no way to be construed to mean that such names may be regarded as unrestricted in respect of trademark and brand protection legislation and could thus be used by anyone.

Cover image: www.ingimage.com

This book is a translation from the original published under ISBN 978-613-9-67834-1.

Publisher:
Sciencia Scripts
is a trademark of
Dodo Books Indian Ocean Ltd. and OmniScriptum S.R.L publishing group

120 High Road, East Finchley, London, N2 9ED, United Kingdom
Str. Armeneasca 28/1, office 1, Chisinau MD-2012, Republic of Moldova, Europe
Printed at: see last page
ISBN: 978-620-8-18145-1

SUMMARY

CHAPTER I

Use of Inoculants in Sugarcane

1. General considerations

The relationship between plants and bacteria can occur in different organs of the plant (roots, stems, leaves), and these microorganisms are called endophytic when they inhabit the interior, and epiphytic when they are in the rhizosphere or on the surface of the tissues. The distribution of these bacteria in plants decreases as they move away from the roots, with the largest populations, in descending order, in the rhizosphere, roots, stem and leaves (SILVA *et al.*, 2012).

Plant-bacteria interaction can contribute positively to plant growth and development, since microorganisms can provide nitrogen, in the process known as biological nitrogen fixation (BNF), growth-promoting hormones such as indole acetic acid (IAA), solubilize phosphates, increase the organic carbon content of the soil, and increase the N content and nutrient retention in the rhizosphere. Thus, bacteria that benefit plants with one or more of the above processes are called growth-promoting bacteria (GPB) (ALIZADEH, 2011; KOCHAR *et al.*, 2011; YADAV *et al.*, 2009).

The bacteria that carry out BNF are called diazotrophs, which have the nitrogenase enzyme capable of breaking the triple bond of atmospheric N_2 and releasing it in the form of NH3 to plants or the soil. Depending on their relationship with plants, these bacteria can be classified in different ways: they can be symbiotic, free-living or associative, the latter being either facultative endophytic or obligate endophytic (BHATTACHARJEE *et al.*, 2008).

The natural occurrence of rhizobia as diazotrophic bacteria in *Fabaceae* has been widely documented (ALVES *et al.*, 2006; GAMARNIK & FRYDMAN, 1991; JESUS *et al.*, 2005; PAULINO *et al.*, 2009; RIBEIRO *et al.*, 2009). Soybeans are a crop in which the practice of nitrogen fertilization has been replaced by the inoculation of seeds with *Bradyrhizobium spp.* which provides the amount of N necessary for its nutrition through BNF (ALVES *et al.*, 2006). However, the presence of diazotrophs is not exclusive to fabaceous plants and has also been observed in non-leguminous plants, such as some *Poaceae*, including sugar cane (BHATTACHARJEE *et al.*, 2008; LIMA *et al.*, 2011; LOIRET *et al.*, 2009; OLIVEIRA *et al.*, 2009).

When compared, the contribution of BNF to nutrients in fabaceous plants is significantly greater than associative BNF in plants such as sugarcane, because symbiotic bacteria (rhizobia) are more efficient in the process than associative bacteria. Taulè *et al.* (2011) observed that the contribution of atmospheric N to sugarcane varied according to the cultivar and species used as a reference plant, from 42 to 59%, compared to 80% in soybeans (ALVES *et al.*, 2006). However, as sugarcane occupies significant tracts of land, this association becomes important (MOREIRA *et al.*, 2010). Global estimates suggest that 16,400,000 tons of N are fixed annually in soybean areas, while in sugarcane areas, the input of N by this route is around 500,000 tons, corresponding to around 0.7 to 1% of the N fixed in agrosystems (HERRIDGE *et al.*, 2008).

2. Brief history and main genera of inoculants for sugar cane

In Brazil, studies on diazotrophic bacteria in grasses began in 1953, with the first isolation of bacteria from the genus *Azotobacter* in acidic soils by Embrapa researcher Johanna Dobereiner, although this genus already had its potential known in Europe and Asia (DOBEREINER, 1953). At the end of the same decade, the bacterium *Beijerinckia fluminensis* was isolated and described in sugarcane rhizosphere, confirming the associative potential of diazotrophic bacteria with grasses (MOREIRA *et al.*, 2010).

Bacteria capable of colonizing the interior of plant tissues and supplying N to plants were described in the 1980s (OLIVEIRA *et al.*, 2009). This was only possible with the creation of semi-solid culture media, where these bacteria can find, in a certain region of the medium, an O diffusion rate$_2$ lower than or equal to their respiration rate, because only in this condition do these bacteria fix N (DOBEREINER *et al.*, 1995; LIMA *et al.*, 2011; MOREIRA *et al.*, 2010).

Azospirilum amazonense was isolated in 1983 in rhizosphere soil and in roots of pupunha, forage grasses, rice, maize and sorghum (DOBEREINER & BALDANI, 1979; DOBEREINER *et al.*, 1995). In 1986, Baldani *et al.* (1986) isolated *Herbaspirilum seropedicae* from maize, rice and sorghum roots, and observed that the isolate showed active nitrogenase enzyme in a semi-solid medium, making it an N-fixing bacterium. Two years later, Cavalcante & Dobereiner (1988) isolated *Gluconacetobacter diazotrophicus* from sugarcane stalks and roots. However, the potential of BNF in sugarcane was only estimated in the following decade by Urquiaga *et al.* (1992), who indicated that 70% of the N accumulated in the biomass came from the atmosphere.

Over the decades, dozens of species of bacteria have been isolated from sugar cane and their abilities have been evaluated, initially in laboratories and then in greenhouse conditions.

The main bacteria isolated from sugarcane tissues (stalk and root) are the species *H. seropedicae*, *H. rubrisubalbicans, G. diazotrphicus, Klebsiella pneumonae, Enerobacter spp, Pantoea sp.*, and other isolates belonging to the genera *Pseudomonas, Kluyera, Staphylococcus, Curtobacterium and Brevibacillus* (BENEDUZI *et al.*, 2013; BHATTACHARJEE *et al.*, 2008; LIMA *et al.*, 2011; LOIRET *et al.*, 2009; SILVA, 2011).

Studies evaluating the inoculation of diazotrophic growth promoting bacteria in sugarcane in Brazil have been led by Embrapa (BALDANI *et al.*, 1986; CAVALCANTE & DOBEREINER, 1988; DOBEREINER, 1953; MOREIRA *et al.*, 2010). After years of isolations, selections and field tests with BPC, Embrapa came to the conclusion that combining bacterial species would be the most viable strategy for improving BNF and crop growth and development. Therefore, a cocktail was formulated with five strains of bacteria BR11335 (*Herbaspirillum seropedicae*), BR11504 (*Herbaspirillum rubrisubalbicans*), BR11281T (*Gluconacetobacter diazotrophicus*), BR11366T (*Burkholderia tropica*) and BR11115 (*Azospirillum amazonense*), which has been evaluated in the field under different soil and climate conditions.

2.1. Characteristics of the bacterial genera in the Embrapa cocktail

2.1.1. Genus *Azospirillum*

The bacteria of the genus *Azospirillum* are endophytic associative diazotrophs, although they can be present in the soil as cysts when there is no host plant. There are around 14 species in this genus, which live in association with mono- and dicotyledonous plants and have a wide ecological distribution. When active, these bacteria can be found in the mucigel of the plant's rhizosphere (external colonization), or they can penetrate and develop in the intercellular spaces (internal colonization). The species *A. amazonense* has been isolated from cereals, forage crops and pupunheira trees in Brazil, showing important growth-promoting mechanisms in sugar cane (MOREIRA *et al.*, 2010).

2.1.2. Genus *Herbaspirillum*

The *Herbaspirillum* genus is made up of around 10 species, not all of which fix N and are associated with plants. The bacteria in this genus occur more rarely than *Azospirillum*. Considered to be obligate endophytes, they have a low survival rate in soils without host crops. However, they are more efficient than other diazotrophs at transferring N to plants. The first species of this genus, *H. seropedica*, was initially classified as an *Azospirillum* because it had a similar growth profile to the genus. The mistake was only undone with the analysis of the 16S rRNA, and a new genus was created. *H. seropedicae* has been isolated from various parts of grasses, colonizing roots, stems and leaves. *H. rubrisulbalbicans* was initially classified as a phytopathogen, causing mottled streak in sugarcane, under the name *Pseudomonas rubrisulbalbicans* (MOREIRA *et al.*, 2010).

2.1.3. Genus *Burkholderia*

Bacteria of the genus *Burkholderia* can bring benefits, such as BPC, or harms to crops, causing diseases and even being pathogenic to humans and animals. This was its first description in plants, still in the genus *Pseudomonas*. The genus is made up of around 62 species, with great functional diversity, from BPC in plants, especially as in BNF, to degrading pollutants. Diazotrophic species can be symbiotic, free-living or even associative, as in the case of the relationship between *B. tropica* and sugar cane. *B. tropica* has also been isolated from the rhizosphere and maize tissues in tropical regions (MOREIRA *et al.*, 2010).

2.1.4. Genus *Gluconacetobacter*

The genus *Gluconacetobacter* is made up of approximately 16 species of endophytic bacteria, four of which have the ability to fix N. Among the fixers, *G. diazotrophicus* belongs to the group of acetic acid bacteria and has been isolated in sugarcane, and this species is credited with making an important contribution to the N fixed by association with the crop (MOREIRA *et al.*, 2010).

2.2. Preparing the inoculant cocktail

The preparation of this inoculant was described by Reis *et al.* (2009) and Oliveira *et al.* (2009), and begins with the cultivation of the pure colonies individually. Each colony is mixed with ground peat (inoculation vehicle), neutralized and sterilized, and packed in 250 g polyethylene bags. These plastic bags are kept at 30°C for seven days for the product to mature, and can be stored at room temperature for 180 days.

The five bags of bacteria are diluted in 300 L of water, so that 450 three-bud corms can be inoculated, whether they are grouped in raffia bags or not. Inoculation is carried out by immersing the sugarcane stalks in a tank containing the diluted cocktail for one hour. After the incubation period, the corms must be planted.

The traditional vehicle used for inoculants has been peat since the 1920s. However, naturally occurring peat in Brazil is of poor quality and restricted to small areas, requiring imports (GUIMARAES *et al.*, 2013). In this sense, various materials have been evaluated, such as sugarcane bagasse, coal dust, rice husks, corn bran and biodegradable polymers, but without much success (SILVA *et al.*, 2009; SILVA *et al.*, 2012).

3. Research into sugarcane inoculants

Muthukumarasamy *et al.* (2006) found that when they inoculated micropropagated sugarcane seedlings *with G. diazotrophicus and Herbaspirillum sp.,* the production of aerial and root biomass was, respectively, 100 and 200% higher than plants that received only one of the bacteria, and 300% higher than the treatment without inoculation, at 45 days after planting. The authors also reported that the N accumulation in the aerial biomass of sugarcane inoculated with the mixture of bacteria was similar to the accumulation shown by plants that received 280 kg ha^{-1} of N. Both treatments showed a 30% higher N accumulation than plants that were inoculated with the bacteria separately, and 100% higher when compared to plants that did not receive nitrogen fertilizer and were not inoculated. However, in this study there was no evidence of BNF and it is suggested that the high dry matter production and high N accumulation promoted by BPC were due to the greater development of the plants, including the root system, which led to better nutrient absorption.

In an eight-month evaluation of the effect of different fertilizers (tanned farmyard manure, NPK, tanned farmyard manure with *Trichoderma viride* and *G. diazotrophicus*) on the development of ratoon sugarcane, Shukla *et al.* (2008) observed that the combination of tanned farmyard manure and bacteria provided greater dry matter production, with an increase of up to 7 t ha^{-1} at harvest, and greater absorption of N, P and K.

In a study carried out on three soils with contrasting chemical and physical attributes, Oliveira *et al.* (2006) observed that inoculation of the sugarcane varieties SP701143 and SP813250 resulted in an increase in BNF, reaching up to 42% of the N absorbed by the crop. In this study, inoculation took place on micropropagated seedlings.

Evaluating inoculant vehicles, Silva *et al.* (2009) found that inoculation of the RB72454 and RB867515 varieties with Embrapa's BPC cocktail promoted an increase in stalk productivity (dry and fresh matter) and N accumulation, with these results being similar to those of canes that received a dose of 120 kg ha^{-1} of N.

Despite the promising results in non-leguminous plants inoculated with BPC, yield increases vary from 0 to 30%, indicating that the optimum conditions for the plant-host association vary according to soil and climate variables and are closely related to the genotype, varying at the species level (OLIVEIRA *et al.*, 2006). Factors such as photosynthetic efficiency, nutritional requirements and resistance to unfavorable conditions, which vary with the genotype, have an important influence on the efficiency of N2 fixation in the relationship between the bacteria and the plants (LIMA *et al.*, 2011).

In this context, it is of fundamental importance to carry out local studies with BPC in sugarcane, isolating, re-inoculating and observing the variables that bring together these genera of bacteria. In a study carried out in five sugarcane-producing municipalities in the state of Rio Grande do Sul, Ben Duzi *et al.* (2013) collected 516 bacterial isolates from the rhizosphere, roots, stalks and leaves of different varieties of sugarcane. However, the number of bacteria that solubilized phosphate and/or produced indolic compounds and/or siderophores varied depending on the municipality and the varieties. Around 290 isolates produced indolic compounds and siderophores, 95 produced indolic compounds and solubilized phosphate, 106 produced siderophores and solubilized phosphate, and 75 showed all three characteristics. In this study, more than 50% of the isolates from the leaves, roots and rhizosphere have not yet been identified in a database using the 16S gene, and the ones that stood out the most were *G. diazotrophicus, Burkholderia sp., g. sp., Achromobacter sp., and Azorhizobium sp.*

Pereira *et al.* (2013) compared nitrogen fertilization (100 kg ha^{-1}) on six sugarcane varieties (RB72454, RB867515, B92579, RB918639, RB92606 and RB855536) with inoculation with Embrapa's cocktail of inoculants and three strains of *Herbaspirillum* sp, BR 11512 (*H. rubrisubalbicans*), BR 11411 (*H. seropedicae*) and BR 11724 (*B. gladioli*). According to the authors, at 210 days after planting, only RB72454 and RB867515 had their stalk yields influenced by the treatments, where RB72454 showed higher yields when inoculated with strain BR 11411, even surpassing nitrogen fertilization, while RB867515 produced more when inoculated with BR 11724.

Schultz *et al.* (2012) observed that the productivity and N accumulation of the RB72454 and RB867515 varieties in the plant-cane cycle did not differ depending on the treatments (no inoculation and no fertilization, only inoculation, and only fertilization) in the municipality of Rio de Janeiro. In the ratoon-cane cycle, when inoculated or fertilized with 120 kg ha^{-1} , the RB867515 variety showed similar production and N accumulation, but was superior to the treatment without fertilization and without inoculation, which shows that the bacteria-plant interaction is dependent on the plant genotype (OLIVEIRA *et al.*, 2006). This result suggests that nitrogen fertilization should be replaced by inoculation with BPC for the RB867515 variety until the first ratoon. According to the authors, this positive response to inoculation was not due to BNF, as the δ^{15} N values of inoculated plants were higher than those of non-inoculated plants, where the opposite effect was expected. The inoculated plants produced more, possibly because they expanded their root system more, having access to a greater reserve of water and nutrients, as a benefit of inoculation with BPC. This benefit is very important for the crop, especially during periods of drought, when the plants are subjected to water stress, as was the case in this study. Inoculation was carried out with strains of *G. diazotrophicus*, *H. seropedicae*, *H. rubrisubalbicans*, *A. amazonense* and *B. tropica*, previously tested in the laboratory.

In Pernambuco, Silva (2011) isolated bacteria in the rhizosphere, leaves and roots of the RB867515 and RB92579 sugarcane varieties, respectively, the most widely planted in Brazil and in the northeastern states, and also in the RB863129 variety. Most of the bacteria identified in this study are different from those that make up the cocktail indicated by Embrapa, belonging mainly to the genera *Pseudomonas, Pantoea* and *Enterobacter.* However, at least 30% of the isolates in the cane parts were capable of fixing N and solubilizing phosphates, varying according to plant genotype and management, with or without the use of cupinicide (SILVA *et al.,* 2012).

Lima (2012) collected rhizospheric soil and root samples of the sugarcane varieties RB867515 and RB92579, at four and ten months of cultivation of the first ratoon in Pernambuco, to evaluate diazotrophic bacteria. In the first evaluation, the author isolated and identified seven genera of bacteria (*Burkholderia* sp., *Pantoea* sp., *Bacillus* sp., *Erwinia* sp., *Stenotrophomonas* sp., *Enterobacter* sp. and *Pseudomonas* sp.), and in the second evaluation, the same genera were observed, except for *Erwinia* sp., with the genus *Pantoea* sp. being more frequent in RB92579 and *Enterobacter* sp. in RB867515. In this study 94, 100 and 80% of the bacterial isolates were able to fix N, produce AIA and solubilize inorganic phosphate *in vitro.* When the selected isolates were inoculated onto sugarcane stalks in a greenhouse, it was observed that the growth-promoting effects varied according to the plant genotype. The isolate that provided the highest dry and fresh matter production in RB867515 was the one belonging to the *Erwinia* sp. genus, while in RB92579 it was *Burkholderia* sp.

The production of cadaverine, a substance related to plant resistance to abiotic stresses, has been observed in *A. brasilense*. The occurrence of this hormone has been correlated with the formation of adventitious roots in pine, and in the formation of roots in soybeans, in both crops the effects were positive (GAMARNIK & FRYDMAN, 1991; NIEMI *et al.*, 2002). It may also be involved in osmotic regulation, helping to close stomata guard cells (AZIZ *et al.*, 1997). However, this species of bacteria has not been related to sugarcane, but it has been isolated in grasses (CASSAN *et al.*, 2009).

4. Final considerations

The use of BPC in sugarcane farming is an important tool for reducing the use of agricultural inputs without reducing productivity, especially nitrogen fertilizers, either through BNF, by increasing the root system, or by accumulating nutrients in the rhizosphere soil and/or facilitating their absorption. Therefore, with the right combination of variety and inoculants, under certain soil and climate conditions, it will be possible to reduce or omit the application of mineral sources of N, leading to a reduction in costs.

In addition, BPCs can be seen as reducing possible environmental impacts. This is because NH_3 and N_2O (considered an important greenhouse gas) would no longer be emitted into the atmosphere, since much of the N applied to the soil can be lost in these ways (WEIER, 1999), a fact that has not been explored in this review.

5. References

ALIZADEH, O. Effect of Plant Growth Promoting Bacteria on Crop Growth. **American-Eurasian Journal of Sustainable Agriculture**, v. 5, n. 3, p. 344-349, 2011.

ALVES, B. J. R.; ZOTARELLI, L.; FERNANDES, F. M.; *et al.* Biological nitrogen fixation and nitrogen fertilizers in the nitrogen balance of soybean, corn and cotton. **Pesquisa Agropecuària Brasileira**, v. 41, n. 3, p. 449-456, mar 2006.

AZIZ, A.; MARTIN-TANGUY, J.; LARHER, F. Plasticity of polyamine metabolism associated with high osmotic stress in rape leaf discs and with ethylene treatment. **Plant Growth Regulation**, v. 21, p. 153-163, 1997.

BALDANI, J.; BALDANI, V.; SELDIN, L.; DOBEREINER, J. Characterization of Herbaspirillum seropedicae gen. nov., sp. nov., a root-associated nitrogen-fixing bacterium. **International Journal of Systematic Bacteriology**, v. 36, n. 1, p. 86-93, 1986.

BENEDUZI, A.; MOREIRA, F.; COSTA, P. B.; *et al.* Diversity and plant growth promoting evaluation abilities of bacteria isolated from sugarcane cultivated in the South of Brazil. **Applied Soil Ecology**, v. 63, p. 94-104, jan 2013.

BHATTACHARJEE, R. B.; SINGH, A.; MUKHOPADHYAY, S. N. Use of nitrogen-fixing bacteria as biofertiliser for non-legumes: prospects and challenges. **Applied microbiology and biotechnology**, v. 80, n. 2, p. 199-209, aug 2008.

CASSAN, F.; MAIALE, S.; MASCIARELLI, O.; *et al.* Cadaverine production by Azospirillum brasilense and its possible role in plant growth promotion and osmotic stress mitigation. **European Journal of Soil Biology**, v. 45, n. 1, p. 12-19, jan 2009.

CAVALCANTE, V. A.; DOBEREINER, J. A new acid-tolerant nitrogen-fixing bacterium associated with sugarcane. **Plant and Soil**, v. 108, p. 23-31, 1988.

DOBEREINER, J. **Azotobacter in acid soils**. **Bulletin of the Institute of Ecology and Agricultural Experimentation**. Rio de Janeiro: Boletim do Instituto de Ecologia e Experimentaçâo Agricolas. , 1953

DOBEREINER, J.; BALDANI, V. L. Selective infection of maize roots by streptomycin-resistant Azospirillum lipoferum and other bacteria. **Canadian journal of microbiology**, v. 25, n. 11, p. 1264-

9, nov 1979.

DOBEREINER, J.; BALDANI, V. L. D.; BALDANI, J. I. **Como isolar e identificar bactérias diazotróficas de plantas nao-leguminosas**. 1. ed. Itaguai-RJ: Embrapa, 1995. p. 60

GAMARNIK, A; FRYDMAN, R. B. Cadaverine, an Essential Diamine for the Normal Root Development of Germinating Soybean (Glycine max) Seeds. **Plant physiology**, v. 97, n. 2, p. 778-85, Oct 1991.

GUIMARAES, S. L.; BALDANI, V. L. D.; JACOB-NETO, J. Viability of peat inoculant produced with associative bacteria and molybdenum. **Revista Ciência Agronômica**, v. 44, n. 1, p. 10-15, 2013.

HERRIDGE, D. F.; PEOPLES, M. B.; BODDEY, R. M. Global inputs of biological nitrogen fixation in agricultural systems. **Plant and Soil**, v. 311, n. 1-2, p. 1-18, 11 jul 2008.

JESUS, E. DA C.; SCHIAVO, J. A.; FARIA, S. M. Dependence on mycorrhizae for the nodulation of tropical leguminous trees. **Revista Arvore**, v. 29, n. 4, p. 545-552, 2005.

KOCHAR, M.; UPADHYAY, A.; SRIVASTAVA, S. Indole-3-acetic acid biosynthesis in the biocontrol strain Pseudomonas fluorescens Psd and plant growth regulation by hormone overexpression. **Research in microbiology**, v. 162, n. 4, p. 426-35, May 2011.

LIMA, D. R. M. DE. **Nitrogen-fixing bacteria associated with sugarcane plants grown in Pernambuco**. [S.l: s.n.], 2012.

LIMA, R. C. DE; KOZUSNY-ANDREANI, D. I.; ANDREANI JUNIOR, R.; FONSECA, L. Phenotypic Characterization of Endophytic Diazotrophic Bacteria Isolated from Sugar Cane. **Revista Facultad Nacional de Agronomia - Mendellin**, v. 64, n. 1, p. 5803-5813, 2011.

LOIRET, F. G.; GRIMM, B.; HAJIREZAEI, M. R.; KLEINER, D.; ORTEGA, E. Inoculation of sugarcane with Pantoea sp. increases amino acid contents in shoot tissues; serine, alanine, glutamine and asparagine permit concomitantly ammonium excretion and nitrogenase activity of the bacterium. **Journal of plant physiology**, v. 166, n. 11, p. 1152-61, 15 jul 2009.

MELO, S. R. DE; ZILLI, J. É. Biological nitrogen fixation in cowpea cultivars recommended for the state of Roraima. **Pesquisa Agropecuâria Brasileira**, v. 44, n. 9, p. 1177-1183, 2009.

MOREIRA, F. M. DE S.; SILVA, K.; NÓBREGA, R. S. A.; CARVALHO, F. DE. Associative diazotrophic bacteria: diversity, ecology and potential applications. **Comunicata Scientiae**, v. 1, n. 2, p. 74-99, 2010.

MUTHUKUMARASAMY, R.; GOVINDARAJAN, M.; VADIVELU, M.; REVATHI, G. N-fertilizer saving by the inoculation of Gluconacetobacter diazotrophicus and Herbaspirillum sp. in micropropagated sugarcane plants. **Microbiological research**, v. 161, n. 3, p. 238-45, jan 2006.

NIEMI, K.; HAGGMAN, H.; SARJALA, T. Effects of exogenous diamines on the interaction between ectomycorrhizal fungi and adventitious root formation in Scots pine in vitro. **Tree physiology**, v. 22,

n. 6, p. 373-81, Apr 2002.

OLIVEIRA, A. L. M.; STOFFELS, M.; SCHMID, M.; *et al.* Colonization of sugarcane plantlets by mixed inoculations with diazotrophic bacteria. **European Journal of Soil Biology**, v. 45, n. 1, p. 106-113, jan 2009.

OLIVEIRA, A. L. M.; CANUTO, E. L.; URQUIAGA, S.; REIS, V. M.; BALDANI, J. I. Yield of micropropagated sugarcane varieties in different soil types following inoculation with diazotrophic bacteria. **Plant and Soil**, v. 284, n. 1-2, p. 23-32, jun 2006.

PAULINO, G. M.; ALVES, B. J. R.; BARROSO, D. G.; URQUIAGA, S.; ESPINDOLA, J. A. A. Biological nitrogen fixation and transfer by leguminosae in organic mango and soursop orchards. **Pesquisa Agropecuâria Brasileira**, v. 44, n. 12, p. 1598-1607, dec 2009.

PEREIRA, W.; LEITE, J. M.; HIPÓLITO, G. DE S.; SANTOS, C. L. R. DOS; REIS, V. M. Biomass accumulation in sugarcane varieties inoculated with different strains of diazotrophic bacteria. **Revista Ciência Agronômica**, v. 44, n. 2, p. 363-370, 2013.

REIS, V. M.; PEREIRA, W.; HIPÓLITO, G. DE S. **Technical Communication 118 - Methods for applying diazotrophic bacteria to sugarcane plants to determine agronomic efficiency**. . [S.l: s.n.]. , 2009

SCHULTZ, N.; MORAIS, R. F. DE; SILVA, J. A. DA; *et al.* Agronomic evaluation of sugarcane varieties inoculated with diazotrophic bacteria and fertilized with nitrogen. **Pesquisa Agropecuâria Brasileira**, v. 47, n. 1, p. 261-268, 2012.

SHUKLA, S. K.; YADAV, R. L.; SUMAN, A.; SINGH, P. N. Improving rhizospheric environment and sugarcane ratoon yield through bioagents amended farm yard manure in udic ustochrept soil. **Soil and Tillage Research**, v. 99, n. 2, p. 158-168, jun 2008.

SILVA, M. D. E. O. **BACTERIA ASSOCIATED WITH SUGAR CANE: ISOLATION AND PLANT GROWTH PROMOTION POTENTIAL**. [S.i: s.n.], 2011.

SILVA, M. D. O.; FREIRE, F. J.; LIRA JÛNIOR, M. A.; *et al.* Isolation and prospecting of endophytic and epiphytic bacteria in sugarcane in areas with and without terminicide. **Revista Brasileira de Ciência do Solo**, v. 36, n. 4, p. 1113-1121,2012.

SILVA, M. F. DA; ANTÔNIO, C. DE S.; OLIVEIRA, P. J. DE; *et al.* Survival of endophytic bacteria in polymer-based inoculants and efficiency of their application to sugarcane. **Plant and soil**, v. 356, p. 231-243, 2012.

SILVA, M. F. DA; OLIVEIRA, P. J. DE; XAVIER, G. R.; RUMJANEK, N. G.; REIS, V. M. Inoculants formulated with polymers and endophytic bacteria for sugarcane cultivation. **Pesquisa Agropecuària Brasileira**, v. 44, n. 11, p. 1437-1443, nov 2009.

TAULÉ, C.; MAREQUE, C.; BARLOCCO, C.; *et al.* The contribution of nitrogen fixation to sugarcane

(Saccharum officinarum L.), and the identification and characterization of part of the associated diazotrophic bacterial community. **Plant and Soil**, v. 356, n. 1-2, p. 35-49, 26 Oct 2011.

WEIER, K. L. N2O and CH4 emission and CH4 consumption in a sugarcane soil after variation in nitrogen and water application. **Soil Biology and Biochemistry**, v. 31, p. 1931-1941, 1999.

YADAV, R. L.; SUMAN, A.; PRASAD, S. R.; PRAKASH, O. Effect of Gluconacetobacter diazotrophicus and Trichoderma viride on soil health, yield and N-economy of sugarcane cultivation under subtropical climatic conditions of India. **European Journal of Agronomy**, v. 30, n. 4, p. 296-303, May 2009.

CHAPTER II

N Cycle in Sugarcane Crops in the Plant-Cane Cycle, with Harvesting Without Shredding by Fire - Annual Estimates

1. General considerations

This cycle of N in the soil-plant-atmosphere system in the sugarcane crop (Figure 1) was developed by compiling various national and international publications. The data on productivity and N allocation in the different parts of the sugarcane was based on studies that used 60 kg ha^{-1} of N, and when possible, the application of 200 g ha^{-1} of Mo as fertilizer management, in order to be similar to our experiment. Thus, we considered a yield (stalk) of 101 t ha^{-1} , and N accumulations of 98, 24.5, 27.5, 9.3, 12.4, 72.3 kg ha^{-1} , respectively, in the stalks, leaves, tops, underground stalks, roots and leaves deposited on the ground (OLIVEIRA, 2012; SALCEDO & SAMPAIO, 1984).

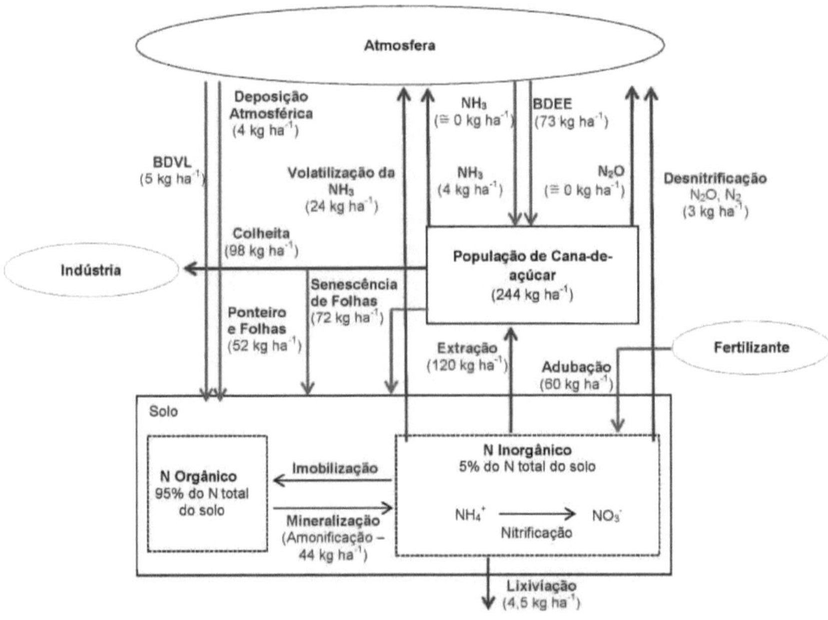

Figure 1. N cycle in sugarcane cultivation without fire stripping, with annual estimates. *The ellipses represent the clouds, the rectangles the compartments and the arrows the flows (red - outflows; blue - inflows; black - internal transformations) of N in the system.

2. Nitrogen inputs, outputs and flows in the soil-plant system

N is the most abundant chemical element in the atmosphere, accounting for around 78% of gases, with N predominating$_2$. However, due to the triple bond between its two atoms, this form of N is very stable and is not accessible to most organisms, plants and animals (FIELDS, 2004). In this case, atmospheric N can enter the soil-plant system, represented in the cycle as soil-sugarcane population, by three processes: atmospheric deposition, fertilization (industrial fixation) and biological fixation (BOUWMAN *et al.*, 2005).

Atmospheric deposition occurs through rainfall or even dust and lightning strikes, and is called wet deposition and dry deposition respectively, whereby NH_3 is mainly deposited in the soil. In agricultural environments with an unpolluted atmosphere, the value of this input varies from 3 to 5 kg ha^{-1} of N (BOUWMAN *et al.*, 2005).

The supply of nutrients to crops through inorganic or organic inputs is called fertilization. The most common form of nitrogen fertilization for sugar cane is mineral inputs such as urea and ammonium sulphate. These fertilizers are obtained through industrial fixation, with the fixation of N_2 to soluble salts, which increase the availability of N for plants when applied to the soil. This fixation can take place in three ways: by direct oxidation, by the cyanamide process and by ammonia synthesis, the latter of which produces 95% of soluble fertilizers (HIREL *et al.*, 2011; SMITH, 2002). The recommended nitrogen fertilizer for sugar cane in Pernambuco is 40 kg ha^{-1} (IPA, 2008), however, the crop responds well in terms of productivity to a dose of 60 kg ha^{-1} (OLIVEIRA, 2012), the dose used in the cycle.

Organic matter (OM) is a potential source of N for plants, because the nutrient is found in organic forms and needs to be converted into inorganic forms for absorption. The OM already present in the soil can provide the crop with N, even in the plant-cane cycle (FRANCO *et al.*, 2011). In general, the mineralization of organic N in the soil is around 8% per year (CAMARGO *et al.*, 1997).

Considering that the soil has 1.15 and 0.20 dag kg^{-1} , respectively, of total organic carbon and total N, and a density of 1.36 g cm^{-3} (OLIVEIRA, 2012; RHODEN *et al.*, 2006), the mineralization of N mentioned above provides 44 kg ha $.^{-1}$

In sugarcane plantations, the main source of OM applied is filter cake, which contains around 1% N (SANTOS *et al.*, 2011). However, this practice was not considered in the N cycle in sugarcane cultivation presented here, since producers don't have enough of this organic input to apply to the entire sugarcane plantation, and only apply it to small areas.

The other way in which N2 enters the soil or plants is through biological fixation (BNF), promoted by bacteria that possess the nitrogenase enzyme and are capable of breaking the triple bond of atmospheric N at room temperature and pressure (ALVES *et al.*, 2006; BARBER *et al.*, 1976; HERRIDGE *et al.*, 2008). These diazotrophic bacteria can live in the soil without a host, i.e. free-living (BDVL), or in association with plants, however, in both cases N is fixed in the form of NH3 (BOUWMAN *et al.*, 2005). These associative bacteria can inhabit the external or internal tissues of sugarcane, and are called epiphytic and endophytic, respectively. In general, BDVL deposit around 5 kg ha^{-1} of N, while BDEE in sugarcane fix up to 70% of the N absorbed by the crop, however, for the cycle the average value of 30% was chosen (BODDEY *et al.*, 2003; BOUWMAN *et al.*, 2005;

TAULÉ *et al.*, 2011).

In the soil, NH_3 from BNF is reduced to NH_4 +, which together with NO_3^- and NO_2^- , makes up the inorganic phase of N in the soil. However, N-NO2$^-$ is an ephemeral form and is practically not absorbed by plants. Under aerobic conditions, NH_4^+ , whether it comes from BNF, atmospheric deposition, mineralization of OM or fertilizers, can be: oxidized by bacteria of the genus *Nitrosomonas* and *Nitrospira* to NO2$^-$, which in turn is oxidized to NO3$^-$ by bacteria of the genus *Nitrobacter*; immobilized in the microbial biomass if C is available; absorbed by plants; lost through leaching or ammonia volatilization. When immobilized in the microbial biomass, N-NH_4 + becomes part of amino acids and proteins, and when these microorganisms die, they release the absorbed N into the soil (ARMAS *et al.*, 1992; GARGANTINI & CATANI, 1957; HIREL *et al.*, 2011; HOGBERG, 1997; TRIVELIN *et al.*, 1998).

On the other hand, N-NO3$^-$, whether it comes from fertilizers or from the nitrification of NH4$^+$ can be: absorbed by the sugar cane; immobilized in the microbial biomass; leave the soil-plant system through leaching and denitrification, the latter being lost to the atmosphere in the form of gases, N_2 and/or N_2 O. Denitrification occurs when there is a lack of O_2 in the soil and microorganisms use N oxides as electron receptors. This is a condition that occurs in waterlogged soils, or occasionally in aggregates saturated with water, reaching a value of 3 kg ha^{-1} of N (BOUWMAN *et al.*, 2005; GARGANTINI & CATANI, 1957; GAVA *et al.*, 2003; HOGBERG, 1997; OLIVEIRA *et al.*, 1999; ROBINSON *et al.*, 2011).

NO3$^-$ is the main form of N lost through leaching. In the cane-plant cycle, this form of N that leaves the reach of the root system derives mainly from the soil and crop remains, around 4.5 kg ha^{-1} , with no N losses from fertilizers (OLIVEIRA *et al.*, 1999).

The "balance" between organic and inorganic N is achieved through the activity of the soil microbiota, which account for 95 and 5% of total N, respectively. Inorganic forms of N in the soil, whether they come from fertilizers or the mineralization of organic matter, are quickly incorporated into the microbial biomass and become unavailable to plants, or simply become immobilized. As part of the microorganism population dies, it releases the N into the soil, still in organic form, which can be mineralized by other microorganisms (CAMARGO *et al.*, 1997; RHODEN *et al.*, 2006). The mineralization of N can be illustrated by the reaction below, where R and E mean, respectively, an organic radical and the energy released in the reaction (CANTARELLA, 2007):

$$\text{Proteína} + H_2O \xrightarrow{\text{protease}} R\text{-}NH_2 + CO_2 + E + \text{outros produtos}$$

$$R\text{-}NH_2 + H_2O \xrightarrow{\text{Aminoácido desidrogenase}} R + NH_3 + E$$

$$NH_3 + H^+ \longleftrightarrow NH_4^+$$

Plants can exchange NH_3 with the atmosphere through their leaves. Depending on the direction of the concentration gradient, the leaves can lose or gain N-NH_3 . However, it is more common for the gas concentration to be higher in the leaves, in which case the plant tissue loses N to the atmosphere. In general, this loss varies from 2 to 6 kg ha^{-1} in annual crops (HOLTAN-HARTWIG & BOCKMAN, 1994; TRIVELIN *et al.*, 2002).

Plants can also lose N as N2O through their leaves. When plants absorb NO_3^- from the soil, in order to use this form of N in amino acids and proteins, they need to reduce it to ammonium, and in this process there can be losses through denitrification. In this case, this happens first by the enzyme nitrate reductase, which reduces it to nitrite, and then by nitrite reductase, after which glutamine synthetase produces glutamine, which is incorporated by glutamate synthetase into amino acids. However, these losses are small and difficult to measure (CANTARELLA, 2007).

When raw sugarcane is harvested, normally only the stalks are removed from the field and sent to the agro-industries for the production of alcohol and/or sugar, leaving the leaves, tops, roots and underground stalks, which could contribute to the supply of N in the next cycle (TIVELIN *et al.*, 2002). In addition, during the crop cycle the old leaves tend to senesce and be deposited on the ground, or follow this flow through the shredding process at harvest time. Estimates of the deposition of dry leaves in the soil indicate 14 t ha^{-1} , providing 72 kg ha^{-1} of N (SALCEDO & SAMPAIO, 1984).

3. References

ALVES, B. J.R.; ZOTARELLI, L.; FERNANDES, F. M.; *et al.* Biological nitrogen fixation and nitrogen fertilizers in the nitrogen balance of soybean, corn and cotton. **Pesquisa Agropecuâria Brasileira**, v. 41, n. 3, p. 449-456, mar 2006.

AMAS,R. DE; VALADIER, M. H.; CHAMPIGNY, M. .; LAMAZE, T. Influence of Ammonium and Nitrate on the Growth and Photosynthesis of Sugarcane. **Journal of Plant Physiology**, v. 140, p. 531-535, 1992.

BARBER, L. E.; TJEPKEMA, J. D.; RUSSELL, S. A.; EVANS, H. J. Acetylene Reduction (Nitrogen Fixation) Associated with Corn Inoculated with Spirillum. **Applied Environmental of Microbiology**, v. 32, n. 1, p. 108-113, 1976.

BODDEY, R. M.; URQUIAGA, S.; ALVES, B. J. R.; REIS, V. Endophytic nitrogen fixation in sugarcane: present knowledge and future applications. **Plant and Soil**, v. 252, n. 1, p. 139-149, May 2003.

BOUWMAN, A. F.; DRECHT, G. VAN; HOEK, K. W. VAN DER. Surface N balances and reactive N loss to the environment from global intensive agricultural production systems for the period 1970 - 2030. **Science in China Series C Life Sciences**, v. 48, p. 1-13, 2005.

CAMARGO, F. A. O.; GIANELLO, C.; VIDOR, C. NITROGEN MINERALIZATION POTENTIAL IN SOILS OF RIO GRANDE DO SUL. **Revista Brasileira de Ciência do Solo**, v. 21, n. 1, p. 575-579, 1997.

FIELDS, S. Nitrogen Global Cycling out of Control. **Environmental Health Perspectives**, v. 112, n.

10, p. 556-563, 2004.

FRANCO, H. C. J.; OTTO, R.; FARONI, C. E.; *et al.* Nitrogen in sugarcane derived from fertilizer under Brazilian field conditions. **Field Crops Research**, v. 121, n. 1, p. 29-41, feb 2011.

GARGANTINI, H.; CATANI, R. A. **DETERMINATION OF THE TIME PERIOD FOR AMMONIFICATION AND NITRIFICATION OF VARIOUS NITROGEN FERTILIZERS**. **Technical Bulletin**. Campinas-SP: Instituto Agronòmico do Estado de Sâo Paulo. , 1957

GAVA, G. J. C.; TRIVELIN, P. C. O.; VITTI, A. C.; OLIVEIRA, M. W. NITROGENIUM (15N) RECOVERY FROM URINE AND FLEET BY SUGAR CANE SOILS (Saccharum spp.). **Revista Brasileira de Ciência do Solo**, v. 27, p. 621-630, 2003.

HERRIDGE, D. F.; PEOPLES, M. B.; BODDEY, R. M. Global inputs of biological nitrogen fixation in agricultural systems. **Plant and Soil**, v. 311, n. 1-2, p. 1-18, 11 jul 2008.

HIREL, B.; TÉTU, T.; LEA, P. J.; DUBOIS, F. Improving Nitrogen Use Efficiency in Crops for Sustainable Agriculture. **Sustainability**, v. 3, n. 12, p. 1452-1485, 7 sep 2011.

HOGBERG, P. 15N natural abundance in soil-plant systems. **New Phytologist**, v. 137, p. 179203, 1997.

OLIVEIRA, A. C. DE. **Interaction between nitrogen and molybdic fertilization in sugar cane Thesis submitted to the Agronomy (Soil Science) Postgraduate Program of the University of São Paulo.**

Federal Rural University of Pernambuco , as part of the requirements for obtaining the degree of Doctor of Agriculture [S.l.]: UFRPE, 2012.

OLIVEIRA, M. W.; TRIVELIN, P. C. O.; VITTI, G. J. C. G. A. C. Nitrogen Leaching in Soil Cultivated with Sugarcane: Lysimeter Experiment. **STAB**, v. 18, n. 2, p. 28-31, 1999.

RHODEN, A. C.; SILVA, L. S. DA; BRITZKE, D.; RANNO, S. K. RELATIONSHIP BETWEEN VARIED SOIL CHARACTERISTICS AND NITROGEN ABSORPTION BY IRRIGATED RICE IN TWO SUCCESSIVE CROPS IN A VARIED HOUSEHOLD. **Revista Brasileira de Agrociências**, v. 12, n. 3, p. 325-331, 2006.

ROBINSON, N.; BRACKIN, R.; VINALL, K.; *et al.* Nitrate paradigm does not hold up for sugarcane. **PloS one**, v. 6, n. 4, p. e19045, jan 2011.

SALCEDO, I. H.; SAMPAIO, E. V. S. B. **Utilization efficiency of -15N urea by sugarcane in the coastal tableland of Pernambuco**. **Regional Workshop on Nuclear Techniques in Crop Production**. Piracicaba-SP: [s.n.]. , 1984

SANTOS, D. H.; SILVA, M. D. A.; TIRITAN, C. S.; FOLONI, J. S. S.; ECHER, F. R. Technological quality of sugarcane under fertilization with filter cake enriched with soluble phosphate Technological quality of sugarcane under fertilization with filter cake enriched with soluble phosphate. **Revista**

Brasileira de Engenharia Agricola e Ambiental, v. 15, n. 5, p. 443-449, 2011.

SMITH, B. E. Structure. Nitrogenase reveals its inner secrets. **Science (New York, N.Y.)**, v. 297, n. 5587, p. 1654-5, 6 sep 2002.

TAULÉ, C.; MAREQUE, C.; BARLOCCO, C.; *et al.* The contribution of nitrogen fixation to sugarcane (Saccharum officinarum L.), and the identification and characterization of part of the associated diazotrophic bacterial community. **Plant and Soil**, v. 356, n. 1-2, p. 35-49, 26 Oct 2011.

TRIVELIN, P. C. O.; BENDASSOLLI, J. A.; MURAOKA, M. W. O. T. Potential of Mixing Aquamonia with Vinasse in Fertilizing Sugarcane Harvested Without Fire Stripping. Part II: Losses through Ammonia Volatilization and Recovery of N Applied to the Soil. **STAB**, v. 16, n. 3, p. 23-26, 1998.

TRIVELIN, P. C. O.; OLIVEIRA, M. W.; VITTI, A. C.; GAVA, G. J. C.; BENDASSOLLI, J. A. Nitrogen losses from urea in the soil-plant system in two sugarcane cycles. **Pesquisa Agropecuària Brasileira**, v. 37, n. 2, p. 193-201, 2002.

TRIVELIN, P. C. O.; VITTI, A. C.; OLIVEIRA, M. W.; GAVA, G. J. C.; SARRIÉS, G. A. Nitrogen utilization and sugarcane productivity in sandy soil with incorporation of crop residues. **Revista Brasileira de Ciência do Solo**, v. 26, n. 3, p. 637-646, 2002.

CHAPTER III

Techniques for Evaluating Biological Nitrogen Fixation

Techniques for evaluating biological nitrogen fixation (BNF) by free-living, associative or symbiotic bacteria in relation to plants allow prior knowledge of their potential in the laboratory and their confirmation and quantification in greenhouse and/or field experiments. In the laboratory, tests are carried out with diazotrophic bacteria isolated and purified from plant tissues and soil. The isolates can be tested qualitatively in a medium without N or by the activity of nitrogenase (AN), an essential enzyme for fixation. In field or greenhouse work, in addition to AN, BNF can be estimated by methods that use the N balance in the soil-plant system; by using the stable isotope^{15} N, and also by correlating it with nitrogen compounds related to fixation (BODDEY *et al.*, 2000; CHALK, 1985; DOBEREINER *et al.*, 1995; HARDY *et al.*, 1968; HERRIDGE *et al.*, 2008; HOGBERG, 1997).

1. Techniques for identifying N-fixing bacteria

In laboratory conditions, BNF has been indicated by isolating the bacteria from plant tissues or soil and growing them *in vitro* in N-free medium. When you want to find out about possible endophytic diazotrophic bacteria, you should first collect the material and wash it in running water to remove excess impurities. To treat roots, fragments of the tissue should be sterilized sequentially with 70% alcohol, sodium hypochlorite, 70% alcohol and water. When the plant tissue is leaves and stems, after washing with running water and drying with paper, surface sterilization is carried out with cotton soaked in 70% alcohol. And when you want to isolate bacteria from seeds, you can treat them with 95% ethanol, acidified hypochlorite and water (DOBEREINER *et al.*, 1995; SILVA, 2011).

After disinfection, the plant tissues are macerated in a solution of PBS and then the mixture is shaken for 40 minutes at 120 rpm. The extract can then be diluted and applied to N-free semi-solid culture media. Culture media can vary according to the species of bacteria, such as JNFb for *Herbaspirillum* spp and semi-solid LGI-P for *Acetobacter diazotrophicus*. Normally, the colonies formed are repeated one or more times in order to completely isolate the desired bacterial colonies, and in some cases the selective potato medium can be used (DOBEREINER *et al.*, 1995; SILVA, 2011).

When you want to obtain N-fixing bacteria in the soil, you use samples of soil sieved through very fine sieve or even soil from the rhizosphere, both of which must be air-dried. About 0.1 g of soil is spread on plates containing the growth medium for a certain period, which can vary according to the bacterial species. The incubation medium varies according to the bacterial species; for *Beijerinckia* spp. the most suitable medium is Silica Gel, while for *Derxia* spp. and *Azospirillum* spp. LGD and semi-solid NFb are recommended respectively (DOBEREINER *et al.*, 1995; SILVA, 2011).

Once the isolates have been obtained, they are transferred to a medium selective for diazotrophic bacteria and incubated for a certain period, depending on the bacterial species and the growth medium containing bromothymol blue (DOBEREINER *et al.*, 1995). Confirmation of BNF is given by bacterial growth in the form of a clear halo and a change in the color of the medium (Figure 1B) due to a change in pH, as shown in the figure below:

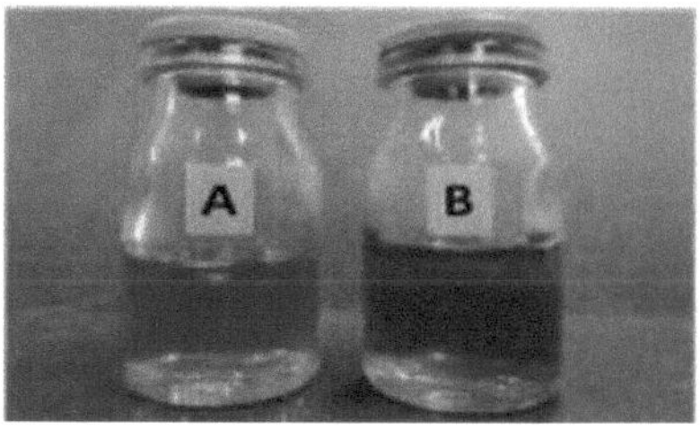

Figure 1: *In vitro* biological nitrogen fixation test in NFb culture medium. A: negative control, UAGC 980 (*Dyella* sp.), bacterial growth without the formation of a bacterial halo; B: positive test carried out with isolate UAGC 857 (*Burkholderia* sp.). Adapted from Lima (2012).

When the aim is to evaluate BNF by symbiotic bacteria in faba beans, i.e. N-fixing bacteria, these bacteria must be isolated, multiplied *in vitro* and inoculated into plants, *in* order to compare the development of these plants with non-inoculated plants and/or plants without and/or with N (HUNGRIA & ARAÙJO, 1994).

Another way of qualitatively assessing BNF in bacterial isolates is by estimating AN using the acetylene reduction method (HARDY *et al.*, 1968), which is described below in section 2.1.

2. Techniques for estimating BNF in plants

In work with plants, in the greenhouse or in the field, the capacity for BNF by bacteria has been indicated or even estimated, mainly by determining the activity of the nitrogenase enzyme and by methods that use the stable isotope^{15} N. Other methods can also be applied, such as the simple N balance, the total N balance and the ureide balance. However, they are heavily weighted, which makes them not very useful and/or usable.

2.1. Acetylene reduction test

The most widely used technique for assessing AN is known as acetylene reduction. This technique is based on the principle that when nitrogenase is in the presence of acetylene (C2H2), it reduces it to ethylene (C2H4), and this happens with the same energy expenditure as the conversion of N2 to NH3. Therefore, the rate at which acetylene is reduced is indicated as the AN, and can be determined in bacteria in the soil, roots, stems, leaves, and even in isolates in Petri dishes (HARDY *et al.*, 1968).

In early research into BNF, AN was used to estimate the contribution of N2 to plant nutrients. In general, six electrons are required for the reduction of N2, while two are needed for the reduction of C_2H_2, so for every three moles of C2H2 reduced to C2H4 by nitrogenase, one mole of N2 is reduced to NH3, according to equations 1 and 2. However, the value of this ratio can vary according to the species and the plant's development conditions, such as 3:2 and 4:1 (DOMMERGUES *et al.*, 1973; HARDY *et al.*, 1973; SILVESTER; BENNETT, 1973).

$N2 + 6H^+ + 6e^- \rightarrow 2NH3$ (Equation 1)

$C2H2 + 2H^+ + 2e^- \rightarrow 2C2H4$ (Equation 2)

Using the C2H2:N2 ratio of 3:1, in an experiment conducted in a greenhouse with maize genotypes, (BARBER *et al.*, 1976) observed that AN in roots was 47 nmol h^{-1} g^{-1} , which corresponded to 16 g ha^{-1} day^{-1} of N. For this, the rate of reduced C2H2 is considered constant throughout the day. Yoshida & Ancajas (1973) reported that BNF in flooded rice contributed 52 and 63 kg ha^{-1} of N in the wet and dry seasons, respectively.

Evaluating AN in conifer roots in New Zealand, Silvester & Bennett (1973) observed that the most consistent C2H2:N2 ratio was 3:2. In this case, the reduction of three nmoles of C2H2 corresponded to two nmoles g^{-1} h^{-1} of NH3 or 28 ng g^{-1} h^{-1} of N.

This technique has been widely used in research into BNF because of its high sensitivity, being able to detect the reduction of C2H2 to C2H4 by two to three *Azotobacter* spp. cells; because of the time used in the reduction process, depending on the plant species five minutes is enough to produce a detectable concentration of C H_{24} ; because it uses materials and apparatus that are relatively easy to access; and because it is quick to determine, from one to two minutes (BODDEY *et al.*, 2007).

However, due to the subjectivity of estimating the contribution of N_2 to crop supply by AN, this analysis has been given little importance in BNF studies (JAMES, 2000), and is considered a purely qualitative assessment, indicating whether or not the fixation process has occurred. In this context, techniques based on the ratio of N isotopes (^{14} N and^{15} N) are more recommended for estimating the contribution of BNF to crop nutrition.

2.2. Techniques with the isotope N^{15}

There are two stable isotopes of N,^{14} N and^{15} N, and the concentration of the former is predominant on the planet. The isotope^{15} N occurs in atmospheric N2 at a constant concentration of 0.3663%, which is significantly lower than the levels found in most soils, and this difference is used to estimate BNF (CHALK, 1985; HERRIDGE *et al.*, 2008; PEOPLES *et al.*, 2002).

In theory, a fixing plant has a lower ratio of^{15} N:^{14} N than a non-fixing plant. This is because plants that do not fix N have their requirement for the nutrient met by $NO3^-$ and $NH4^+$ from the soil, which has a higher ratio of^{15} N:^{14} N compared to the atmosphere, which represents the second source of N for fixing plants (CHALK, 1985; HERRIDGE *et al.*, 2008; PEOPLES *et al.*, 2002).

To estimate BNF using knowledge of stable N isotopes, two techniques can be used. When the sources of^{14} N and^{15} N are the soil and atmosphere, the technique is called natural abundance, and isotope dilution, when fertilizers enriched with^{15} N are used, and its disadvantage is its high cost. To determine the proportion of N absorbed by plants that comes from the atmosphere in field experiments, the natural abundance technique of^{15} N is the most suitable (HERRIDGE *et al.*, 2008).

2.2.1. **Natural abundance technique from N^{15}**

In this technique, in addition to the plant in which the fixation is to be known, plants that have no association with N-fixing bacteria are grown as a reference. Therefore, the abundance of ^{15} N in the plant being assessed, which obtains N from the soil and possibly the atmosphere, is compared with a non-fixing plant, which has the soil as its N source, and can be calculated using the following equation (SHEARER & KOHL, 1986):

$$\%Ndfa = 100 \cdot \frac{(\delta^{15}Nref - \delta^{15}Nfix)}{\delta^{15}N\,ref - B} \quad \text{(Equação 3)}$$

Where:

%Ndfa = percentage of N from BNF in the plant under study; δ^{15} *Nref* = natural abundance of ^{15} N in the reference plant;

δ^{15} *Nfix* = natural abundance of ^{15} N in the plant under study;

B = δ^{15} N of the reference plant when totally dependent on N2.

E;

$$\delta^{15}N\ (‰) = \frac{1000 \cdot (^{15}Na - 0{,}3663)}{0{,}3663} \quad \text{(Equação 4)}$$

Where:

^{15}Na = percentage of ^{15} N in the sample.

0.3663 = percentage of ^{15} N in the atmosphere.

Even so, the use of the natural abundance technique of ^{15} N presents considerations that can lead researchers to mistaken conclusions. Among the problems related to this technique are the differences between the plant in which the BNF potential is to be known and the plant used as a reference, since they usually have different cycles, different root systems, exploit different soil depths and may show a temporal difference in N absorption; and the soil, since in highly weathered soils the content of ^{15} N can be as low as in the atmosphere, making it impossible to estimate BNF using this method (BAPTISTA *et al.*, 2014; HOGBERG, 1997). In this way, the isotope dilution technique is more precise and even more accurate than the natural abundance technique.

2.2.2. **Isotope dilution technique from N^{15}**

The concept of this technique is similar to that of the natural abundance of ^{15} N, with the difference that the soil is enriched with the isotope by means of fertilizer. Therefore, this technique can be used in soils with different levels of the isotope, and have the estimate of BNF calculated in a similar way to the natural abundance equation (ASSIS *et al.*, 2002; CHALK, 1985):

$$\%\text{Ndfa} = 100.\frac{(^{15}Nref\ -\ ^{15}Nfix)}{^{15}N\ ref-B}\quad \text{(Equação 5)}$$

Where:

^{15}N *fix* = percentage of atoms of ^{15} N in excess in the plant

2.2.3. *Nref* = percentage of atoms of ^{15} N in excess in the fertilizer

B = percentage of atoms of ^{15} N in excess in the reference plant when totally dependent on N2.

As this technique uses fertilizers enriched with ^{15} N, it has a high cost, which is why it has been used for experiments conducted in containers. However, with adaptations to the size of plots, in micro-plots, BNF has been successfully and economically estimated using this technique (HERRIDGE *et al.*, 2008; PEOPLES *et al.*, 2002).

Just like natural abundance, the isotope dilution technique can also be affected by aspects related to the plant used as a reference (item 2.2). Theoretically, the reference plant cannot be N-fixing and must have a similar cycle and growth of aerial and root biomass, which can lead to an overestimation of BNF (HERRIDGE *et al.*, 2008; PEOPLES *et al.*, 2002).

In this context, it is recommended to grow the fixative plant itself as a reference, in the presence and absence of the enriched fertilizer. In this way, there will be no differences between the reference and fixative plants, and the technique can be used with greater accuracy.

2.3. Simple N

In the simple N balance method, the N accumulated by N-fixing plants is compared with the N accumulated by non-fixing plants grown in the same environment, the difference being considered the N fixed, according to the equation below (HERRIDGE *et al.*, 2008; PEOPLES *et al.*, 2002):

$$N_2fix = Nefix - Nenfix \quad \text{(Equação 6)}$$

Where:

N2fix = N fixed

Nefix = N accumulated by the fixing plant species

Nenfix = N accumulated by the non-fixing plant species

The main premise of this method is that the amount of N absorbed from the soil by fixing and non-fixing plants is similar. This technique can be used successfully in trials with plants that fix considerable amounts of N, grown in soils with low levels of the nutrient. However, there may be differences in the morphology of the root system and/or the ability to exploit the N in the soil between fixing and non-fixing plants, making the method invalid (HERRIDGE *et al.*, 2008; PEOPLES *et al.*, 2002).

2.4. Total N balance method

The total N balance method is based on the principle that the accumulation of N in the soil-plant system over time occurs through the fixation of N_2 . To estimate this, it is necessary to know the inputs (dry and wet deposition, and N gases) and outputs (ammonia volatilization, denitrification, leaching, among others) of N in the system, with BNF being the difference between the N absorbed by the crop and the various nutrient flows. Due to the large number of variables to be controlled, the results of this method are easily underestimated or overestimated, making it a little-used technique (HERRIDGE *et al.*, 2008).

2.5. Ureide method

Ureides are nitrogen compounds produced in the nodules of some N-fixing plants, which are distributed from this tissue to other parts of the plant. Most N-fixing plants export N from the nodules as amides, asparagines and glutamines through the xylem. However, species such as soybeans (*Glycine max*), cowpeas (*Cajanus cajan*), string beans (*Vigna unguiculata*) and snap beans (*Phaseolus vulgaris*), among others, export it mainly as ureides. Therefore, a calibration can be made for these species, where the dependent variable is the relative percentage of ureides in the xylem sap, and the independent variable is the percentage of N derived from the atmosphere. To do this, the plants are grown in N-free substrates and supplied with N in the form of a nutrient solution in increasing concentrations of $NO3^-$, produced with fertilizer enriched with ^{15} N for use in the isotope dilution technique. Once calibrated, BNF can be estimated by measuring the concentration of ureides by spectrophotometry for the plant species (HERRIDGE *et al.*, 1988, 2008; PEOPLES *et al.*, 2002). Thus, after calibration, this technique is simple and low-cost, since the apparatus used to determine ureides is common in soil and plant analysis laboratories.

However, the use of this technique is more efficient in greenhouse experiments and in some plant species.

3. Final considerations

There is a range of methods for indicating and evaluating BNF in bacterial isolates and especially in plants, making it possible to carry out studies in laboratories with simple equipment, such as spectrophotometers, or more complex equipment, such as mass spectrometers. However, the use of more complex and expensive methods, such as the isotope dilution technique of ^{15} N, results in more accurate results. It is therefore up to the researcher, given their financial situation, to choose between the cheapest and least accurate, the most expensive and most accurate, or a combination of methods.

4. References

ASSIS, C. A.; KUBOTA, M.; OHTA, H.; *et al.* Estimation of the nitrogen fixation by sugarcane cultivar NiF-8 using N dilution and natural abundance techniques Estimation of the Nitrogen Fixation by Sugarcane Cultivar NiF-8 Using 15N Dilution and Natural 15 N Abundance Techniques. **Soil**

Science and Plant Nutrition, v. 48, n. 2, p. 283-285, 2002.

BAPTISTA, R. B.; MORAIS, R. F. DE; LEITE, J. M.; *et al.* Variations in the 15N natural abundance of plant-available N with soil depth: Their influence on estimates of contributions of biological N2 fixation to sugar cane. **Applied Soil Ecology**, v. 73, p. 124-129, jan 2014.

BARBER, L. E.; TJEPKEMA, J. D.; RUSSELL, S. A.; EVANS, H. J. Acetylene Reduction (Nitrogen Fixation) Associated with Corn Inoculated with Spirillum. **Applied Environmental of Microbiology**, v. 32, n. 1, p. 108-113, 1976.

BODDEY, L. H.; BODDEY, R. M.; ALVES, B. J. R.; URQUIAGA, S. **The Evaluation of Biological N2 Fixation Associated with Legumes and Non-Legumes Using the Acetylene Reduction Technique: History, Theory and Practice** . Seropédica - RJ: Embrapa Agrobiologia. , 2007

BODDEY, R. M.; PEOPLES, M. B.; PALMER, B.; DART, P. J. Use of the 15 N natural abundance technique to quantify biological nitrogen fixation by woody perennials. **Nutrient Cycling in Agroecosystems**, v. 57, p. 235-270, 2000.

CHALK, M. Estimation of N2 fixation by isotope dilution: an appraisal of techniques involving 15N enrichment and their application. **Soil Biology & Biochemistry**, v. 17, n. 4, p. 389-410, 1985.

DOBEREINER, J.; BALDANI, V. L. D.; BALDANI, J. I. **Como isolar e identificar bactérias diazotróficas de plantas nao-leguminosas**. 1. ed. Itaguai-RJ: Embrapa, 1995. p. 60

DOMMERGUES, Y.; BALANDREAU, J.; RINAUDO, G.; WEINHARD, P. Non-symbiotic nitrogen ftxation in the rhizo-spheres of rice, maize and different tropical grasses. **Soil Biology & Biochemistry**, v. 5, n. Table 1, p. 83-89, 1973.

HARDY, R.; BURNS, R.; HOLSTEN, R. Applications of the acetylene-ethylene assay for measurement of nitrogen fixation. **Soil Biology and Biochemistry**, v. 5, p. 47-81, 1973.

HARDY, R. W.; HOLSTEN, R. D.; JACKSON, E. K.; BURNS, R. C. The acetylene-ethylene assay for n(2) fixation: laboratory and field evaluation. **Plant physiology**, v. 43, n. 8, p. 1185207, aug 1968.

HERRIDGE, D. F.; O'CONNELL, P.; DONNELLY, K. The Xylem Ureide Assay of Nitrogen Fixation : Sampling Procedures and Sources of Error igen Fixation : : of Error. **Journal of Experimental Botany**, v. 39, n. 198, p. 12-22, 1988.

HERRIDGE, D. F.; PEOPLES, M. B.; BODDEY, R. M. Global inputs of biological nitrogen fixation in agricultural systems. **Plant and Soil**, v. 311, n. 1-2, p. 1-18, 11 jul 2008.

HOGBERG, P. 15N natural abundance in soil-plant systems. **New Phytologist**, v. 137, p. 179203, 1997.

HUNGRIA, M.; ARAÙJO, R. S. **Manual of methods used in agricultural microbiology studies.** Embrapa, 1994. 542p.

JAMES, E. K. Nitrogen fixation in endophytic and associative symbiosis. **Field Crops Research**, v.

65, n. December 1999, p. 197-209, 2000.

LIMA, D. R. M. DE. **Nitrogen-fixing bacteria associated with sugarcane plants grown in Pernambuco**. [S.l: s.n.], 2012.

PEOPLES, M. B.; BODDEY, R. M.; HERRIDGE, D. F. Quantification of Nitrogen Fixation. In: LEIGH, G. J. (Ed.). **Nitrogen Fixation at the Millennium**. [S.l.]: Elsevier Science B.V., 2002. p. 357-389.

SHEARER, G.; KOHL, D. H. N2-Fixation in Field Settings: Estimations Based on Natural 15N Abundance. **Australian Journal of Plant Physiology**, v. 13, p. 699-756, 1986.

SILVA, M. D. E. O. **Bacteria associated with sugarcane: isolation and potential for plant growth promotion**. [S.l: s.n.], 2011.

SILVESTER, W. B.; BENNETT, K. J. Acetylene reduction by roots and associated soil of New Zealand conifers. **Soil Biology and Biochemistry**, v. 5, n. 1, p. 171-179, jan 1973.

YOSHIDA, T.; ANCAJAS, R. R. The fixation of atmospheric nitrogen in the rice rhizosphere. **Soil Biology and Biochemistry**, v. 5, n. 1, p. 153-155, jan 1973.

CHAPTER IV

Variation of Biological Nitrogen Fixation by the Availability of Nutrients in the Soil - General Aspects for Legumes and Non-Legumes

1. Initial considerations

Various abiotic factors can interfere with BNF, from the nodulation and inoculation phases of the microorganism to the final processes of fixation. The main factors are soil pH, temperature, water availability and nutrient availability (FIGUEIREDO *et al.*, 2008).

Hungary *et al.* (1985) in an experiment with bean plants in a greenhouse observed that a 25% reduction in light increased nitrogenase activity and the total N accumulated in the aerial part. This effect was attributed to a reduction in maximum daytime temperatures and a reduction in solar energy, which were offset by an increase in leaf area. Temperature generally affects nodulation and the efficiency of nodule activity, with 25°C being the ideal temperature for nitrogenase activity (ZHANG *et al.*1996).

Soil pH is one of the most important factors for BNF, affecting the process both when it is above or below the desirable level, varying according to the plant species (HOWIESON; BALLARD, 2004; BALA; GILLER, 2006 and SHAMSELDIN, 2007). The pH of the soil affects the plant's nutritional relationships by reducing the availability of some nutrients in the soil. In this way, the low availability of nutrients promoted by soil pH leads to a nutritional imbalance in plants, which in turn will reduce root growth and the production of photosimilates, reducing the volume of tissue exposed to colonization and the availability of nutrients for the bacteria that carry out BNF.

Root development can also be affected by high levels of exchangeable aluminum in the soil, reducing rhizobium growth and interfering with root infection (SILVA *et al.* 2010; FAGAN et al. 2007; FIGUEIREDO et *al.* 2008).

Water availability influences the permeability of the nodule membrane to gases, especially oxygen. In plant tissue, water is transported via the symplast and apoplast. Water stress can affect the activity of nitrogenase, limit oxygen absorption in the bacteroid zone by decreasing water flow, or even by low water movement in the phloem, accumulating ureides and aspartate in the nodules and leaves, thus decreasing leghemoglobin synthesis (HUNGRIA; VARGAS 2000; FAGAN *et al.* 2007).

2. Essential elements

The mineral elements available in the soil that are necessary for plant development are called essential elements and are divided into macronutrients and micronutrients (EPSTEIN; BLOOM, 2006).

Macronutrients are essential elements for the plant, needed in greater volume and taking part in its metabolic reactions, without which it cannot complete its life cycle. They are: carbon (C), hydrogen (H), oxygen (O), nitrogen (N), phosphorus (P), sulphur (S), magnesium (Mg), calcium (Ca) and potassium (K) (MALAVOLTA, 2006).

Micronutrients are required in small quantities, from milligrams to micrograms, and also take part in metabolic reactions. The micronutrients are Boron (Bo), Chlorine (Cl), Copper (Cu), Iron (Fe), Manganese (Mn), Molybdenum (Mo), Cobalt (Co), Nickel (Ni) and Zinc (Zn) (MALAVOLTA, 2006). When these elements are absent or in low quantities, they compromise the vitality of plants, as they form part of the structure of amino acids, enzyme complexes and take part in various reactions (Epstein & Bloom, 2006). For this reason, the low availability of nutrients can interfere directly or indirectly with BNF, from nodulation to the final processes of fixation.

3. Elements that interfere with nodulation

Nodulation and infection of the nodule occur simultaneously, the rhizobia that are attached to the root hairs release the Nod factors (lipochitin oligosaccharide signaling molecules), after signaling the rhizobia are enveloped by the root, then there is a degradation of this region forming an infection channel (TAIZ ; ZEIGER, 2004).

The N available in the soil can interfere with root nodulation and BNF. Mineral nitrogen fertilizers are used in the ammoniacal or nitric form; these forms are soluble and available for root absorption. AN is regulated by the ammonia produced by nitrogenase itself inside the nodule. In this way, the amount of fertilizer available can interfere with the enzyme's activity (CAMARA, 2014).

Santos (2014), in an experiment with sugarcane under field conditions, observed that the sugarcane varieties RB867515 and RB92579 fixed N. However, BNF, estimated by the natural abundance method, and AN, by acetylene reduction, decreased as N availability in the soil increased. According to the author, when there is more N available in the soil, plants depend less on BNF, due to energy savings. Since 12 to 17 g of carbohydrate are needed for each N_2 fixed (SANTOS et al. 2008).

The effects of the availability of P, K and S on the growth of plants, especially legumes, and on BNF are interesting both biologically and agronomically, and various direct and indirect effects have been identified. Although many processes remain unclear, attempts are being made to consolidate the fragmentary information on the effects of nutrients on BNF-related traits (DIVITO, 2014).

K acts as an activator of various enzymes in the processes of photosynthesis and respiration and in the activity of nitrogenase. Its absence or deficiency causes a reduction in the dry mass of nodules, thus reducing BNF. Adequate K supply is necessary to sustain acetylene reduction rates and nodule activity by directly activating enzymes, including those involved in ammonia assimilation, amino acid formation, carbon supply and energy transduction (DIVITO, 2014).

Yunis (2010) in an experiment with *Lablab purpureus* (L.) observed that when plants are subjected to water stress, there is a smaller reduction in nodulation and BNF when there is an adequate supply of K, i.e. K increases the plant's tolerance to low water availability.

Ca acts on root growth and chemical signaling proteins (MARSCHNER, 1995). As its absence reduces root growth, there is consequently a reduction in nodulation (CAMARA, 2014). Mg is directly related to photosynthesis (a constituent of chlorophyll) and ATP in oxidative phosphorylation inside the nodule. Its deficiency therefore causes a decrease in BNF (FAGAN *et al* 2007).

S is part of the secondary metabolites that act on nodulation factors, and its deficiency causes a delay or reduction in root nodulation. However, there are few studies evaluating the contribution of these nutrients to nodule fixation (CAMARA, 2014; FAGAN *et al.* 2007; MARSCHNER, 1995).

Biochemical studies have shown a close and positive relationship between S availability and protein content in nodules, particularly nitrogenase and leghemoglobin (VARIN *et al.*, 2010), i.e. with the high proportion of S in the amino acids in both components of nitrogenase, Fe-protein and FeMo-protein (CURATTI *et al.*, 2006).

According to Gordon *et al.* (2001) leghemoglobin maintains a low concentration of free O_2 inside the nodule, so if this protein is found in low concentrations it can result in a higher concentration of O2. Low S availability reduced BNF in peas and lucerne, due to a decrease in ferredoxin and leghemoglobin concentrations, as well as a reduction in ATP supply (SCHERER *et al.*, 2008).

P is directly related to energy production and consumption. When it is supplied in low quantities, there is a reduction in nitrogenase activity and ATP production, consequently reducing nodulation and BNF (SILVA *et al.*, 2010; CAMARA, 2014). Increasing the supply of P promotes an increase in nitrogenase activity and in the accumulation of dry phytomass in the nodule, with high requirements for BNF (FAGAN *et al.*, 2007).

Silva *et al.* (2010) in an experiment carried out in a greenhouse in Piaui, with cowpeas subjected to different doses and sources of soluble P and when evaluating BNF, observed that regardless of the source of P, doses of 60 to 80 kg ha^{-1} led to the greatest crop growth. However, when simple superphosphate was applied, greater nodulation and N fixation was observed in cowpea when compared to triple superphosphate (SILVA *et al.*, 2010).

Some micronutrients also help in the nodulation process, such as boron and nickel (AZEVEDO *et al.* 2002; FRANCO, 2015).

B acts in cell division, which is why when it is not available in the soil it causes a reduction in nodule size (MARSCHNER, 1995). Azevedo *et al.* (2002), when carrying out an experiment in a greenhouse, found that B deficiency affected nitrogenase activity in the nodules of pea plants. In contrast, Campo and Lantmann (1998), in a study carried out in Paranà in a soybean field, found no increase in BNF or productivity following the addition of B. Although it is known that B influences BNF, its role is still little studied (Azevedo *et al.* 2002).

Ni is a constituent of the enzymes urease (hydrolyzes urea) and hydrogenase, which is necessary for reprocessing part of the H generated in the BNF process; this second enzyme is present in bacteria of the genus *Rhizobium* and *Bradyrhizobium*. When Ni is deficient, it delays the nodulation process and reduces BNF (FRANCO, 2015). According to Costa *et al.* (2015), the use of Ni in soybean seeds can, in addition to increasing the efficiency of BNF, provide better use of the reserves contained in the seeds, boosting initial plant growth.

In an experiment carried out in Rio de Janeiro under greenhouse conditions, Campanharo *et al.* (2013) observed that plants grown in Argissolo showed greater growth and higher urea levels in mature leaves when Ni was applied.

Franco (2015) found that applying low doses of Ni (45 and 90 mg kg^{-1}) via soybean seeds increased BNF and favored the accumulation of total N in the grain, with an increase in urease. However, high doses resulted in lower shoot dry mass production and lower grain production.

Copper deficiency decreases BNF. However, its functions are not yet understood (Fagan *et al.* 2007).

Silicon (Si) is a beneficial element for plants. It increases the pH of the soil and can therefore reduce the activity of toxic elements (RODRIGUES *et al.* 2011). Grothge-Lima (1998) found that with high levels of Si in nutrient solution there is a decrease in the development and activity of nodules. Nelwamondo and Dakora (1999), on applying Si, observed a significant increase in the number and mass of *Bradyrhizobium* nodules in cowpea and N fixation.

According to Dakora (2005), cowpea root nodules showed an increase in BNF due to an increase in the number of bacteroids per infected cell, as a result of silicon supplementation.

4. Nutrients involved in reactions catalyzed by nitrogenase

The micronutrients involved in the BNF process through nitrogenase are Mo, Zn, Fe and Mn. Although cobalt is not classified as a micronutrient, but as a beneficial element, it is extremely important for BNF (CAMPO; LANTMANN, 1998).

Mo is absorbed by the plant by mass flow mainly in the form of $MoO2^{4-}$, it is a constituent of the nitrogenase protein molybdate-iron, its deficiency decreases the efficiency of nodulation and causes N deficiency. It is an element that is easily translocated, which is why its deficiency does not occur in younger leaves (TAIZ; ZEIGER, 2004).

Some studies have shown that Mo supplementation in the soil increases BNF. As seen by Campo and Lantmann (1998), the application of Mo to the soil increased the efficiency of BNF.

In a study carried out by Albino and Campo (2001) in a greenhouse with *Bradyrhizobium* strains, it was found that the most efficient Mo sources for BNF are ammonium molybdate, Mo trioxide, Grap and Comol.

In an experiment carried out by Leite *et al.* (2009) under field conditions, it was observed that there was a greater accumulation of N in the aerial part and higher grain yield when Mo was applied to cowpea. Gelain *et al.* (2001) also found that the application of Mo increased yields and protein content in soybeans. This is different to what was observed by Gris *et al.* (2005), who found no increase in soybean productivity when Mo was applied to leaves and grain.

In an experiment carried out in São Paulo under greenhouse conditions, it was found that nodulation in soybeans is affected by the dose and route of application of molybdenum, with a greater number and mass of nodules being obtained with the application of the micronutrient via seed (TOLEDO *et al.*, 2010). Santos (2014) in an experiment carried out in Pernambuco under field conditions observed that the sugarcane varieties RB867515 and RB92579 had BNF, estimated by the natural abundance of^{15} N, and AN was enhanced by molybdenum fertilization, especially in the absence of nitrogen fertilization.

Although MESCHEDE *et al.* (2004) consider molybdic fertilization to be important for BNF, caution should be exercised when applying Mo to the seed, especially in the form of molybdate (MoO), as it tends to reduce the survival of the rhizobium.

Zn acts in the synthesis of leghemoglobin, so when it is deficient, the size of the nodules decreases (MARSCHNER, 1995). Campo and Lantmann (1998) added Zn to the soil, but did not obtain an increase in fixation or productivity in soybeans.

Fe is a constituent of leghemoglobin and the nitrogenase iron-protein, and when deficient it can delay or reduce the start of nodulation as well as reducing BNF (TAIZ; ZEIGER, 2004).

Mn plays a fundamental role in the catalysis of various enzymatic and electron transfer processes. It is a constituent of the enzyme starch hydrolase, which is responsible for the degradation of ureides in the leaves, and in conditions of water stress (drought) the supply of Mn can help maintain BNF (VALDEZ *et al.*, 2000).

Co is a component of the coenzyme cobalamin (B12), a precursor of leghemoglobin, and its deficiency inhibits the synthesis of leghemoglobin. It is absorbed by mass flow mainly in the form of Co^{2+} and its translocation in the plant only occurs after the formation of chelates with organic acids (MARSCHNER, 1995).

Campo and Lantmann (1998) found no increase in nitrogen fixation or in soybean yield after fertilizing with cobalt. This is similar to what was found by Caires and Rosolém (2000) who, when applying Co to peanut seeds, found no effect on nodulation and nitrogen absorption.

Bàrbaro *et al.* (2009) in an experiment carried out in Sâo Paulo under field conditions found that soybeans respond positively to foliar application of Co. This was also confirmed by Godinho *et al.* (2015) in a field experiment on soybeans carried out in Paranà and by Santana *et al.* (2011) who also observed higher grain weights when Co was applied to soybeans regardless of the form of inoculation.

5. Final considerations

Various studies have been carried out on BNF as a function of abiotic factors. However, studies relating nutrient availability and BNF have yet to be carried out. This is because the literature still deals very superficially with the way in which some nutrients alter BNF, despite the fact that they have been shown to do so.

However, in general it can be said that P has a direct relationship with BNF, but specifically with nodulation and the supply of energy to carry out the reactions.

Co, on the other hand, has a directly proportional relationship with fixation, but its specific function in BNF is not yet known, thus suggesting further studies.

B and Ni interfere directly with nodulation.

It is already known that the micronutrients Mo, Zn, Fe and Mn are involved in BNF, however, we suggest more work involving Mn and Zn.

Co and Si, although not micronutrients, contribute to BNF.

Understanding the availability of nutrients in the soil, as well as their role in BNF, is extremely important, since optimizing the fixation process implies higher crop yields and a reduction in the use of nitrogen fertilizers. This in turn reduces economic costs and environmental degradation. For this reason, we stress the need for further research in this area.

6. References

AZEVEDO, W.R.; FAQUIN, V.; MOREIRA, F.M.S.; OLIVEIRA JÙNIOR,A.C.; LISBOA, C.C. Efeito do boro na nodulaçao da ervilha cultivada em solos de vàrzea. **Pesquisa agropecuària brasileira,** Brasilia, v. 37, p. 1137-1143, 2002.

CAIRES, E.F. ; ROSOLEM, C.A. Nodulation and nitrogen absorption by peanut in response to liming, cobalt and molybdenum. **Scientia agricola**. v.57, p.337-341,2000.

CAMARA, G.M.S.; Biological nitrogen fixation in soybeans, **International plant nutrition institute -Brazil**, p.9, 2014.

CAMPANHARO, M.; MONNERAT P.H.; ESPINDULA, M.C.; RABELLO, W.S. Nickel doses in cowpea grown in two soils. **Revista Caatinga**, v. 26, p. 10 - 18, 2013.

CURATTI, L., LUDDEN, P.W., RUBIO, L.M. Nif B-dependent in Vitrosynthesis of the ironmolybdenum cofactor of nitrogenase. **Proceedings of the National Academy of Sciences**. U.S.A. v. 104, p. 5207-5301, 2006.

DIVITO, G.A.; SADRAS, V.O. How do phosphorus, potassium and sulphur affect plant growth and biological nitrogen fixation in crop and pasture legumes? A meta-analysis. **Field Crops Research**. v. 156, p.161-171, 2014.

EPSTEIN, E.; BLOOM, A. **Nutriçâo mineral de Plantas: Principios e Perspectivas**. 2 ed. Londrina: Planta, 2006.402 f.

FAGAN, E.B.; MEDEIROS, S. L. P.; MANFRON, P.A; CASAROLI, D.; SIMON, J.; NETO, D.D.; LIER, Q.J.V.; SANTOS, O.S.; MÜLLER, L. Physiology of biological nitrogen fixation in soybeans -

REVIEW. **Revista da FZVA**. Uruguaiana, v.14, p. 89-106, 2007.

FIGUEIREDO, M.V.B.; LIRA JÙNIOR, M. A.; ARAÙJO, A. S. I.; MARTINEZ, C. R. Biotic and abiotic factors for biological N2 fixation. In: FIGUEIREDO, M.V.B.; BURITY,H. A.; STAMFORD, N. P.; SANTOS, C. E. R. **Agrobiodiversity: the new challenge for agriculture. Agrolivros**, 2008, 568p.

FRANCO, G.C. Treatment of soybean seeds with nickel to increase biological fixation and urease activity. **Dissertation** (Master's Degree - Science Postgraduate Program. Area of Concentration: Chemistry in Agriculture and the Environment). Center for Nuclear Energy in Agriculture, University of Sâo Paulo. Piracicaba, 64 p, 2015.

GODINHO, F.B.; SILVA LIMA, L.H.; FERRI, G.C.; BRACCINI, A.L.;PIETRO, J.P.C.; BIANCHESSI, P.A. Evaluation of different formulations and dosages of cobalt and molybdenum on soybean productivity. **Anais.** IX EPCC - Encontro Internacional de Produçâo Cientifica UniCesumar. p. 4-8, 2015.

GRIS, E. P.; CONTE E CASTRO, A.M. ; OLIVEIRA, F.F. Soybean productivity in response to molybdenum application and inoculation with Bradyrhizobium japonicum. **Revista Brasileira de Ciência do Solo,** v. 29, p.151-155, 2005.

GROTHGE-LIMA, M.T. Interrelationship between stem canker (Diaporthe phaseolorum f. sp. meridionalis), nodulation (Bradyrhizobium japonicum) and silicon in soybeans [Glycine max (l.) merrill]. 1998. 58p. **Thesis** (Doctorate), Center for Nuclear Energy in Agriculture, University of Sâo Paulo. Piracicaba - SP.

MARSCHNER, H. Nitrogen Fixation. In: MARSCHNER, H. (Ed.) **Mineral nutrition of higher plants**. 2 ed. London: Academic Press, 1995. P. 201-228.

RODRIGUES, F.A.; OLIVEIRA, A.L.; KORNDORFER, A.P.; KORNDORFER, G.H. "Silicio: Um elemento benèfico e importante para plantas". **Informaçoes Agronômicas**. p.14-20. 2011.

SANTANA, M.J. SOUSA, F.C.A.; SILVEIRA,A.L.;SILVA, C.A. Aplicação de cobalto, molibdênio e inoculante na cultura da soja (*glycine max l. merrill*). **Global science and technology.** v. 04, p.01 - 08, 2011.

SANTOS, C.E.R.S.; FREITAS, A.D.S.;VIEIRA, I.M.M.B.; et al. Symbiotic N2 fixation in tropical legumes. In FIGUEIREDO, M.V.B.; BURITY,H. A.; STAMFORD, N. P.; SANTOS, C. E. R. **Microorganisms and Agrobiodiversity: the new challenge for agriculture.** Guaiba: Agrolivros, 2008. p. 17- 41.

SANTOS, R.L. **Molybdenum in metabolism and biological nitrogen fixation in sugarcane.** 2014. 135f. Thesis (PhD in Soil Science) Federal Rural University of Pernambuco - Recife, PE.

SILVA, E.F.L; ARAÙJO, A.S.F; SANTOS, V. B. ; NUNES, L.A.P.L.;CARNEIRO, R.F.V. Biological fixation of N^2 in cowpea under different doses and sources of soluble phosphorus. **Bioscience Journal.** Uberlândia, v. 26, p. 394-402, 2010.

TAIZ, L. ; ZEIGER, e. **Plant Physiology**. 3°ed, Porto Alegre: Artmed, 2004, 722 p.

TOLEDO, M.Z.; GARCIA, R.A.; PEREIRA, M.R.R.; BOARO, C.F.S.; G.P.P. LIMA. Nodulation and nitrate reductase activity as a function of molybdenum application in soybean. **Bioscience Journal.** v. 26, p. 858-864, 2010.

VALDEZ, D.; SINCLAIR, T. R., SERRAJ, R.; PURCELL, L. C. Manganese application alleviates the water deficit-induced decline of N2 fixation. **Plant Cell and Environment**, v.23, p.497-505, 2000.

CHAPTER V

Biological Fixation and Activity of Enzymes Related to Nitrogen Assimilation in Sugarcane under the Effect of Nitrogen and Molybdic Fertilization

1. Initial considerations

Brazil is a traditional sugarcane grower, being the largest producer, responsible for 33% of world production (JADOSKI *et al.*, 2010). In addition to the country's high sugarcane production, there is also a nutritional aspect that draws attention. Compared to other sugarcane producers such as India, the USA, Colombia and Australia, nitrogen fertilization in Brazil is still low, at an average of 40% of the doses applied in these countries (BAPTISTA *et al.*, 2014).

Several studies have left a gap in relation to the origin of the N absorbed by sugarcane, and a range of authors can be found in the literature who suggest that the highest percentage of N comes from non-mineral sources. Umesh *et al.* (2013), in India under field conditions, observed sugarcane extracting around 100 kg ha^{-1} of N, however, only 75 kg ha^{-1} were applied, a deficit of 25 kg ha^{-1} , in relation to what was incorporated into the soil, for a yield of 47 t ha^{-1} without considering system losses, which can exceed 50% of the fertilizer applied (BOUWMAN *et al.*, 2005).

Using nitrogen fertilizer enriched with^{15} N, Franco *et al.*, 2011 observed that throughout the development of the crop, its contribution to the N absorbed varied from 15 to 44% at 134 after planting to 5 to 13% at harvest, in the cane-plant cycle. In the ratoon, the unidentified N source was more important, contributing around 80% of the total N. In both cycles, the authors observed that the higher the dose of nitrogen fertilizer, the greater the amount of N absorbed from this source, but the unidentified source always contributed more to the crop's nutrition.

Knowing that N in its inorganic form in the soil comprises only 2% of total N, that its concentration is low in relation to crop demand, being expressed in mg kg^{-1} , and that the contribution of nitrogen fertilizer to the crop cycle is low, it can be said that most of the nutrient absorbed by sugarcane comes from the mineralization of soil organic matter and/or the atmosphere (BAPTISTA *et al.*, 2014; BOUWMAN *et al.*, 2005; FRANCO *et al.*, 2011). These facts have supported the hypothesis that the crop may be associated with bacteria that perform biological nitrogen fixation (BNF), which can provide significant amounts of N to the crop, which tends to reduce the use of N-source inputs (OLIVEIRA *et al.*, 2011).

In the global nitrogen cycle, atmospheric N enters the soil without the presence or use of electrical discharges and high temperatures, carried out through organisms, which may or may not be associated with plants, that possess the enzyme called nitrogenase, capable of reducing N2 to NH3 (HU *et al.*, 2011). Every year, it is estimated that the input of N through biological fixation associated with sugarcane reaches 500,000 tons, corresponding to around 0.7 to 1% of the N fixed in agrosystems. However, global estimates of BNF in sugarcane are still unreliable due to the variation in values found by researchers around the world. In Brazil, the 43

BNF contributes around 10 to 20% of the total N, but there are estimates higher than this (HERRIDGE *et al.*, 2008).

2. Effect of molybdenum on biological nitrogen fixation

Nitrogenase is an enzyme made up of two units, one reductive and the other catalytic. The reducing unit, called dinitrogenase reductase or Fe-protein, is responsible for transferring electrons to the reduction of N_2 , while in the reaction site, the dinitrogenase, the reduction to NH3 actually takes place. Dinitrogenase can have three compositions, MoFe-protein, VFe-protein and FeFe-protein, depending on the availability of metals in the environment. However, in environments where all three metals are available, dinitrogenase composed of Mo, V and Fe is preferred in decreasing order, i.e. the nitrogenase unit with V is only formed when Mo is no longer available, and the one with Fe when Mo and V are not available (HERNANDEZ *et al.*, 2008; HU *et al.*, 2011; WEARE *et al.*, 2006; YOSHIZAWA *et al.*, 2010).

The possible variations in the second unit of nitrogenase are also differentiated by their efficiency in the use of energy in the process of producing NH3 from N2, with the most efficient set of reactions taking place through MoFe-protein (GUHA & PHUKAN, 2011), as shown in equations 1 and 2. In this context, changes in the components that make up the nitrogenase enzyme can affect its activity, such as Fe and Mo.

Under optimum conditions, a temperature of 30° C and a saturating quantity of N2, the reduction of N2 to NH3 by Mo and V nitrogenase takes place according to equations 1 and 2 respectively:

$N_2 + 8e^- + 8H^+ + 16MgATP \rightarrow 2NH_3 + H_2 + 16MgADP + 16P_i$ (Equation 1)

$N_2 + 12e^- + 12H^+ + 24MgATP \rightarrow 2NH_3 + 3H_2 + 24MgADP + 24P_i$ (Equation 2)

As can be seen, the production of two moles of NH_3 by VFe-protein requires 50% more energy than the conversion carried out by MoFe-protein.

The most widely used technique for assessing nitrogenase (AN) activity is known as acetylene reduction. When nitrogenase is in the presence of acetylene (C2H2), it reduces it to ethylene (C2H4), and this happens with the same energy expenditure as the conversion of N2 to NH3. Therefore, the rate at which acetylene is reduced is indicated as AN (HARDY *et al.*, 1968).

AN is reduced or null when there is a high availability of inorganic N in the soil, as a mechanism of energy regulation by the plants. As the reduction of N2 to NH_3 involves energy expenditure for the plants, if there is enough N available for crop nutrition, the host will not reduce the supply of photosynthesized compounds to the bacteria. This is evidenced in the work carried out by Day *et al.* (1975), who observed that up to seven hours after the application of nitrogen fertilizer, AN in *Paspalum notatum* roots was reduced, and after 24 hours the enzyme's activity returned to values close to those presented by plants with a continuous supply of the nutrient. The authors also found that *Brachiaria decumbens* and *Pennisetum purpureum* showed reduced AN two weeks after the application of 20 kg ha^{-1} of N, when compared to the values of plants that did not receive the nutrient.

After Fe, Mo is the most common metal in dinitrogenase and is an enzyme regulator. As such, its concentration in the medium and incorporation into MoFe-protein are directly linked to AN and are directly proportional. This is evidenced *in the* work carried out *in vitro* by Hernandez *et al.* (2008), who, when synthesizing nitrogenase, observed that ethylene production, i.e. AN, was increased as the concentration of Mo in the medium and its incorporation into dinitrogenase increased.

Li-Ping *et al.* (2007a) evaluated the behavior of AN in two sugarcane varieties (RB72454 and RB867575) under the effect of four doses of Mo (0; 0.01; 0.02 and 0.04 mg L^{-1}), in the absence and presence of N. According to the authors, under nitrogen fertilization the doses of Mo promoted an increase in AN in the roots and stalk of the varieties, and in the leaves only in RB72454. In the absence of N, the doses of Mo increased AN in the leaves and stalk of the RB867575 variety, but reduced the activity of the enzyme in the leaves, stalk and roots of RB72454, which shows the difference in behavior between varieties, requiring studies for each genotype.

In addition to its regulatory effect on AN, Mo is a micronutrient for plants that is normally found in the soil in low levels. Generally, the low availability of Mo, which is observed in acidic soils (pH < 5.0) with a high presence of Fe and Al oxides, leads crops grown on them to be deficient in Mo (BRENNAN & BOLLAND, 2007; FITRI *et al.*, 2008). Raising the pH 3 of a soil by one unit can increase the availability of Mo by 100 times, so it is necessary to correct acidic soils with limestone in order to increase its availability (KAISER *et al.*, 2005).

In areas where applying limestone was unfeasible, Brennan & Bolland (2007) observed that wheat yields were increased by applying Mo to soils with an average pH of 4.3 in Australia. The increase in grain yield in these soils was only obtained with an increase of up to 100% of the recommended dose of the micronutrient, which was applied to the seeds and top-dressed.

As has been reported, BNF in plants can be indicated by the AN of the bacteria associated with them. However, this analysis is predominantly qualitative and does not indicate, in a more precise way, the value of the N absorbed by the crops from the atmosphere, requiring more accurate techniques to assess this contribution, in this case for sugar cane.

To estimate the contribution of BNF to the supply of nutrients to crops, the most recommended techniques are those based on the proportion of N isotopes (^{14} N and^{15} N), which can be either natural abundance, when the sources of^{14} N and^{15} N are practically the soil and atmosphere, or isotopic dilution using fertilizers enriched with^{15} N, the disadvantage of which is the high cost. To determine the proportion of N absorbed by plants that comes from the atmosphere in field experiments, the natural abundance technique of^{15} N is the most suitable (HERRIDGE et al., 2008). Even so, the use of the natural abundance of^{15} N technique has some drawbacks that can lead researchers to erroneous conclusions. Among the problems related to the^{15} N natural abundance technique is the difference between the plant in which the BNF potential is to be known and the plant used as a reference, since they usually have different cycles and exploit different soil depths (BAPTISTA *et al.,* 2014).

However, both methods have demonstrated the contribution of N2 to sugarcane nutrition. Assis *et al.* (2002) estimated the contribution of BNF in sugarcane using the techniques of natural abundance and isotopic dilution of^{15} N. According to the authors, using the natural abundance technique, at six months after planting, in pots containing soil, 32% of the N absorbed by the crop came from the atmosphere. Corroborating the existence of BNF, the authors cultivated sugarcane in nutrient solution and observed at five months after cultivation, using the isotope dilution technique, that there was BNF, where around 26, 41 and 21% of the N contained in the roots, stalks and leaves, respectively, came from the atmosphere.

Using the natural abundance technique of^{15} N, Taulè *et al.* (2011) observed that the contribution of atmospheric N to sugarcane varied according to the cultivar and species used as a reference plant, from 42 to 59%. The authors also reported that BNF decreased by 15% as the dose of N increased, from 10 to 50 mg kg^{-1} of N. The experiment was conducted in Uruguay with national varieties, in pots with soil enriched with the isotope^{15} N, in a greenhouse until 120 days after planting.

However, studies on the effect of N and Mo fertilization on BNF and enzymatic activities are more advanced in the soybean crop, providing indications for future research with other crops. In a study carried out in pots containing soil in a greenhouse, Sun *et al.* (2012) observed that BNF and RNA in soybean rhizosphere soil was higher in the presence of Mo when compared to soils without the micronutrient.

Campo *et al.* (2009) found that soybeans responded to molybdic fertilization with an increase in productivity, with a "supposed" increase in BNF, when evaluating plants grown from seeds with different Mo contents. When plants produced from soybean seeds rich in Mo were inoculated with *Bradyrhizobium* and then given more Mo, either in the seed or via foliar fertilization, they showed similar productivity and N accumulation to plants produced from seeds with a low Mo content, inoculated with the same bacteria and given 200 kg ha^{-1} of N via mineral fertilizer. This compensation for nitrogen fertilization by molybdic fertilization was possibly due to the increase in AN in the bacteria. Therefore, when soybeans are produced with Mo supplementation, the seeds have a higher content of the micronutrient, which leads to a residual effect for the high productivity of the next crop, confirming its fundamental importance in BNF.

The interaction between BNF and the application of nitrogen fertilizers tends to be driven by energy expenditure by crops. According to Salvagiotti *et al.* (2008), although soybeans respond to N application by increasing productivity, the use of N-source inputs reduces BNF, due to the energy savings that plants grown with fertilizer have. The authors observed that the behavior of BNF in soybeans under increasing doses of N has a negative exponential relationship, i.e. tending to zero as the dose of N increases.

3. **Effect of molybdenum on nitrate reductase activity**

In plants, this metal is a component of the nitrate reductase enzyme, which is responsible for reducing nitrate to nitrite in the N-NO3 assimilation pathway$^-$ (FITRI *et al.*, 2008). The catalytic flux or activity of nitrate reductase (NRA) can be influenced, among other things, by the availability of $NO3^-$ and Mo, which are directly proportional (KAISER *et al.*, 2005).

When a plant is deficient in Mo, its N metabolism is altered, especially when the predominant form of N in the soil is nitrate. Once absorbed, nitrate can be stored in vacuoles or reduced to nitrite in the root plastids or leaf chloroplasts by the enzyme nitrate reductase, whose component is Mo, to form amino acids and proteins (RUFTY *et al.*, 1986). Thus, when a plant is deficient in Mo, the formation and activity of the enzyme is reduced or null, and the first stage of the reduction of nitrate to ammonium is not initiated, i.e. there is a reduction in N assimilation, with a consequent reduction in growth and productivity (KAISER *et al.*, 2005).

In an experiment conducted with sugarcane varieties in a hydroponic system with sand, Li-Ping *et al.* (2007) observed that RNA was higher when the plants were grown in the presence of N, up to five months after planting. The authors also reported that the activity of the enzyme was increased by increasing the concentration of Mo in the nutrient solution; however, the increases were neither proportional nor constant, both in the presence and absence of N.

Jongruaysup *et al.* (1994) observed that when india beans *(Vigna mungo L.)* received a dose of 0.64 mg kg^{-1} of Mo, N accumulation was increased by up to 1,000% when compared to the control, i.e. plants that did not receive Mo, at 45 days after sowing. This was possibly due to the increase in RNA driven by Mo. The trial was carried out in pots containing sandy loam soil, pH 4.3, in Australia.

Weng *et al.* (2009) evaluated the effect of increasing doses of Mo on *Chamaecrista rotundifolia*, a perennial herbaceous *Fabaceae*. According to the authors, NA, ARN and dry matter production were affected by the doses of the micronutrient. In general, the enzyme activities increased as the dose of Mo increased, with the greatest differences being observed between the treatment that did not receive Mo and the treatments that did. However, the AN values in the control treatment were close to the other treatments, suggesting that the Mo content in the soil, 0.22 mg kg^{-1} , was sufficient to meet the enzyme's demand. On the other hand, the NRA in the plants that didn't receive Mo was different from the other plants, indicating that the pre-existing level of the nutrient in the soil wasn't enough to meet the demand for the reduction of nitrate to nitrite in the N-NO3 assimilation pathway$^-$. Dry matter production behaved similarly to NRA, which suggests that if *C. rotundifolia* is to be grown in soils with Mo levels lower than or equal to 0.22 mg kg^{-1} , it needs to be supplemented with fertilizers.

In addition to the availability of $NO3^-$ and Mo mentioned above, NRA varies between sugarcane genotypes (HEMAPRABHA *et al.*, 2013). Among other things, the preference for absorbing one of the forms of N, N-$NO3^-$ or $NH4^+$, alters the enzyme's activity. The greater the presence of nitrate in the soil and its absorption, which is more common (WANG *et al.*, 2011), the greater the activity of the enzyme in the plant. If the genotype prefers ammonium, the enzyme's activity is low and may be zero.

In the experiment carried out by Armas *et al.* (1992), it was observed that when sugar cane was subjected to an environment with N-NH_4 + it grew more, showing greater height, greater stem diameter, longer and wider leaves, which led to greater aerial biomass production. However, the RNA in the leaves of these plants was reduced, representing only 45% of the activity of the plants grown with N-$NO3^-$. The evaluations were carried out on six-month-old plants that had been receiving a nutrient solution with N in cationic or anionic form for 45 days, in a substrate composed of vermiculite and peat.

In a study carried out in a hydroponic system using fertilizers enriched with^{15} N (^{15} NH4NO3 and/or NH4^{15} NO3), Robinson *et al.* (2011) observed that when sugarcane varieties had contact with a solution containing a high concentration of N with similar proportions of nitrate and ammonium for 24 hours, they absorbed more of the cationic form of the nutrient. According to the authors, the preference in the incorporation of ammonium to nitrate was around five and four to ten times, respectively, in the roots and tillers.

The aforementioned preference for ammonium over nitrate by sugarcane varieties was reinforced by field experiments. After opening a trench and incubating the root system *in situ* for 30 and 120 minutes in a nutrient solution containing fertilizer enriched with^{15} N, the authors also observed that regardless of the incubation time, N absorption in the cationic form was higher than in the anionic form. It is likely that in both cases the NRA was low (ROBINSON *et al.*, 2011).

4. Final considerations

In view of the above, BNF in sugarcane can be affected by the application of Mo and N, but some caveats should be made about research on the subject. BNF in different species and their cultivars has commonly been assessed using isotopic techniques, but little is known about the effect of Mo and the N*Mo interaction on fixation, especially in sugarcane, making it a research niche of isolated knowledge. The situation becomes even more abstract when we look for studies evaluating the behavior of the N*Mo interaction on BNF in sugarcane, which are scarce or even non-existent.

There is at least a consensus in the scientific community that BNF is reduced by the greater availability of N in the soil. Since the association with bacteria takes place at an energy cost to the plants, if N availability is high, there will be no fixation.

In trials evaluating the effect of Mo, fixation is only indicated by AN, which, as already mentioned, is estimated by the reduction in acetylene, and can only be a temporary qualitative analysis of possible BNF. NRA, N absorption and assimilation and productivity are still evaluated as a function of Mo, in the hope that they will be increased, and some positive results have been found, but the experiments are not usually reproduced in the field.

But in general, it is believed that it is possible that Mo increases NRA, causing plants to assimilate N more quickly and thus absorb more of the nutrient and increase productivity. However, if N absorption and assimilation increase, BNF could be reduced. And even if the RNA is not altered by Mo, the AN is increased, increasing the BNF.

5. References

ARMAS, R. DE; VALADIER, M. H.; CHAMPIGNY, M. .; LAMAZE, T. Influence of Ammonium and Nitrate on the Growth and Photosynthesis of Sugarcane. **Journal of Plant Physiology**, v. 140, p. 531-535, 1992.

ASSIS, C. A.; KUBOTA, M.; OHTA, H.; *et al.* Estimation of the nitrogen fixation by sugarcane cultivar NiF-8 using N dilution and natural abundance techniques Estimation of the Nitrogen Fixation by Sugarcane Cultivar NiF-8 Using 15N Dilution and Natural 15 N Abundance Techniques. **Soil Science and Plant Nutrition**, v. 48, n. 2, p. 283-285, 2002.

BAPTISTA, R. B.; MORAIS, R. F. DE; LEITE, J. M.; *et al.* Variations in the 15N natural abundance of plant-available N with soil depth: Their influence on estimates of contributions of biological N2 fixation to sugar cane. **Applied Soil Ecology**, v. 73, p. 124-129, jan 2014.

BOUWMAN, A. F.; DRECHT, G. VAN; HOEK, K. W. VAN DER. Surface N balances and reactive N loss to the environment from global intensive agricultural production systems for the period 1970 - 2030. **Science in China Series C Life Sciences**, v. 48, p. 1-13, 2005.

BRENNAN, R. F.; BOLLAND, M. D. A. Increased Concentration of Molybdenum in Sown Wheat Seed Decreases Grain Yield Responses to Applied Molybdenum Fertilizer in Naturally Acidic Sandplain Soils. **Journal of Plant Nutrition**, v. 30, n. 12, p. 2005-2019, 3 dec 2007.

CAMPO, R. J.; ARAUJO, R. S.; HUNGRIA, M. Molybdenum-enriched soybean seeds enhance N accumulation, seed yield, and seed protein content in Brazil. **Field Crops Research**, v. 110, n. 3, p. 219-224, feb 2009.

DAY, J. M.; NEVES, M. C. P.; DOBEREINER, J. NITROGENASE ACTIVITY ON THE ROOTS OF TROPICAL FORAGE GRASSES. **Soil Biology & Biochemistry**, v. 7, p. 107-112, 1975.

FITRI, N.; KASTENHOLZ, B.; BUCHARI; AMRAN, M. B.; WARGANEGARA, F. M. Molybdenum Speciation in Raw Phloem Sap of Castor Bean. **Analytical Letters**, v. 41, n. 10, p. 1773-1784, 16 jul 2008.

FRANCO, H. C. J.; OTTO, R.; FARONI, C. E.; *et al.* Nitrogen in sugarcane derived from fertilizer under Brazilian field conditions. **Field Crops Research**, v. 121, n. 1, p. 29-41, feb 2011.

GUHA, A. K.; PHUKAN, A. K. Why vanadium complexes perform poorly in comparison to related molybdenum complexes in the catalytic reduction of dinitrogen to ammonia (Schrock cycle): a theoretical study. **Inorganic chemistry**, v. 50, n. 18, p. 8826-33, 19 sep 2011.

HARDY, R. W.; HOLSTEN, R. D.; JACKSON, E. K.; BURNS, R. C. The acetylene-ethylene assay for n(2) fixation: laboratory and field evaluation. **Plant physiology**, v. 43, n. 8, p. 1185207, aug 1968.

HEMAPRABHA, G.; SIMON, S.; LAVANYA, D. L.; SAJITHA, B.; VENKATARAMANA, S. Evaluation of Drought Tolerance Potential of Elite Genotypes and Progenies of Sugarcane (Saccharum sp . hybrids). **Sugar Tech**, v. 15, n. 1, p. 9-16, 2013.

HERNANDEZ, J. A; CURATTI, L.; AZNAR, C. P.; *et al.* Metal trafficking for nitrogen fixation: NifQ donates molybdenum to NifEN/NifH for the biosynthesis of the nitrogenase FeMo-cofactor. **Proceedings of the National Academy of Sciences of the United States of America**, v. 105, n. 33, p. 11679-84, Aug 19, 2008.

HERRIDGE, D. F.; PEOPLES, M. B.; BODDEY, R. M. Global inputs of biological nitrogen fixation in agricultural systems. **Plant and Soil**, v. 311, n. 1-2, p. 1-18, 11 jul 2008.

HU, Y.; LEE, C. C.; RIBBE, M. W. Extending the carbon chain: hydrocarbon formation catalyzed by vanadium/molybdenum nitrogenases. **Science (New York, N.Y.)**, v. 333, n. 6043, p. 753-5, Aug 5, 2011.

JADOSKI, C. J.; TOPPA, E. V. B.; JULIANETTI, A.; *et al.* Physiology of vegetative stage development of sugarcane (Saccharum officinarum l.). **Pesquisa Aplicada & Agrotecnologia**, v. 3, n. 2, p. 169-176, 2010.

JONGRUAYSUP, S.; DELL, B.; BELL, R. W. Distribution and Redistribution of Molybdenum in Black Gram (Vigna mungo L. Hepper) in Relation to Molybdenum Supply. **Annals of Botany**, v. 73, p. 161-167, 1994.

KAISER, B. N.; GRIDLEY, K. L.; BRADY, J. N.; PHILLIPS, T.; TYERMAN, S. D. The Role of Molybdenum in Agricultural Plant Production. **Annals of Botany**, v. 96, p. 745-754, 2005.

LI-PING, W.; YANG-RUI, L.; LI-TAO, Y. Effects of molybdenum on nitrogenase activity of nitrogen-fixing bacteria in sugarcane.pdf. **Plant Nutrition and Fertilizer Science**, v. 13, n. 1, p. 110-122, 2007a.

LI-PING, W.; YANG-RUI, L.; LI-TAO, Y. Effects of Molybdenum on Nitrogen Metabolism of Sugarcane. **Sugar Rec**, v. 9, n. 1, p. 36-42, 2007b.

OLIVEIRA, Z. M. DE; FLOH, E. I. S.; FERRARA, F. I. S.; BARBOSA, H. R. Diazotrophyc rhizobacteria isolated from sugarcane can release amino acids in a synthetic culture medium. **Biology and Fertility of Soils**, v. 47, n. 8, p. 957-962, 2011.

ROBINSON, N.; BRACKIN, R.; VINALL, K.; *et al.* Nitrate paradigm does not hold up for sugarcane. **PloS one**, v. 6, n. 4, p. e19045, jan 2011.

RUFTY, T. W.; THOMAS, J. F.; REMMLER, J. L.; CAMPBELL, W. H.; VOLK, R. J. Intercellular localization of nitrate reductase in roots. **Plant physiology**, v. 82, n. 3, p. 675-80, nov 1986.

SALVAGIOTTI, F.; CASSMAN, K. G.; SPECHT, J. E.; *et al.* Nitrogen uptake, fixation and response to fertilizer N in soybeans: A review. **Field Crops Research**, v. 108, n. 1, p. 1-13, jul 2008.

SUN, T.; WANG, Y. P.; WANG, Z. Y.; LIU, P.; XU, G. D. The effects of molybdenum and boron on the rhizosphere microorganisms and soil enzyme activities of soybean. **Acta Physiologiae Plantarum**, v. 35, n. 3, p. 763-770, 29 sep 2012.

TAULÉ, C.; MAREQUE, C.; BARLOCCO, C.; *et al.* The contribution of nitrogen fixation to sugarcane (Saccharum officinarum L.), and the identification and characterization of part of the associated diazotrophic bacterial community. **Plant and Soil**, v. 356, n. 1-2, p. 35-49, 26 Oct 2011.

UMESH, U. N.; KUMAR, V.; ALAM, M.; SINHA, S. K.; VERMA, K. Integrated Effect of Organic and Inorganic Fertilizers on Yield, Quality Parameter and Nutrient Availability of Sugarcane in Calcareous Soil. **Sugar Tech**, v. 15, n. 4, p. 365-369, Apr 23, 2013.

WANG, T.-H.; CHEN, Y.-H.; HUANG, J.-Y.; *et al.* Enzyme kinetics, inhibitors, mutagenesis and electron paramagnetic resonance analysis of dual-affinity nitrate reductase in unicellular N(2)- fixing cyanobacterium Cyanothece sp. PCC 8801. **Plant physiology and biochemistry : PPB / Société française de physiologie végétale**, v. 49, n. 11, p. 1369-76, nov 2011.

WEARE, W. W.; DAI, X.; BYRNES, M. J.; *et al.* Catalytic reduction of dinitrogen to ammonia at a single molybdenum center. **Proceedings of the National Academy of Sciences of the United States of America**, v. 103, n. 46, p. 17099-106, Nov 14, 2006.

WENG, B.-Q.; HUANG, D.-F.; XIONG, D.-Z.; *et al.* Effects of Molybdenum Application on Plant Growth, Molybdoenzyme Activity and Mesophyll Cell Ultrastructure of Round Leaf Cassia in Red Soil. **Journal of Plant Nutrition**, v. 32, n. 11, p. 1941-1955, 9 oct 2009.

YOSHIZAWA, J. M.; FAY, A. W.; LEE, C. C.; HU, Y.; RIBBE, M. W. Insertion of heterometals into the NifEN-associated iron-molybdenum cofactor precursor. **Journal of biological inorganic chemistry : JBIC : a publication of the Society of Biological Inorganic Chemistry**, v. 15, n. 3, p. 421-8, mar 2010.

MIX
Papier aus verantwortungsvollen Quellen
Paper from responsible sources
FSC® C105338

Printed by Books on Demand GmbH, Norderstedt / Germany